Ricin Toxins

Ricin Toxins

Special Issue Editor

Nilgun E. Tumer

MDPI • Basel • Beijing • Wuhan • Barcelona • Belgrade

Special Issue Editor
Nilgun E. Tumer
University of New Jersey
USA

Editorial Office
MDPI
St. Alban-Anlage 66
4052 Basel, Switzerland

This is a reprint of articles from the Special Issue published online in the open access journal *Toxins* (ISSN 2072-6651) from 2018 to 2020 (available at: https://www.mdpi.com/journal/toxins/special issues/ricin toxins).

For citation purposes, cite each article independently as indicated on the article page online and as indicated below:

LastName, A.A.; LastName, B.B.; LastName, C.C. Article Title. *Journal Name* **Year**, *Article Number*, Page Range.

ISBN 978-3-03928-512-9 (Hbk)
ISBN 978-3-03928-513-6 (PDF)

Contents

About the Special Issue Editor

Nilgun E. Tumer is a Distinguished Professor at the Department of Plant Biology at Rutgers, the State University of New Jersey. She obtained her Ph.D. in biochemistry from Purdue University and trained with Dr. Robert Haselkorn at the University of Chicago. Before coming to Rutgers, she worked at Monsanto Company as a group leader in virus resistance. The major focus of her laboratory is to understand the basic mechanism of toxicity of ribosome inactivating proteins (RIPs), such as ricin and Shiga toxins (Stxs) and trichothecene mycotoxins produced by *Fusarium graminearum*. Dr. Tumer's lab pioneered yeast as a powerful model to study the mode of action of RIPs. Her research played a central role in defining the ribosomal targets of RIPs by investigating the interactions of ricin and Stxs with ribosomes using surface plasmon resonance (SPR) with Biacore instruments. They developed the protocols for screening fragment and small molecule libraries to identify inhibitors for ricin and Stxs. Dr. Tumer has published 120 papers and has 15 issued patents. She directs the Core Facility at the School of Environmental and Biological Sciences (SEBS), which provides a wide array of services to the Rutgers community.

Editorial

Introduction to the *Toxins* Special Issue "Ricin Toxins"

Nilgun E. Tumer

Department of Plant Biology, School of Environmental and Biological Sciences, Rutgers University, New Brunswick, NJ 08901-8520, USA; tumer@sebs.rutgers.edu; Tel.: +1-848-932-6359

Received: 20 December 2019; Accepted: 24 December 2019; Published: 27 December 2019

Ricin toxin isolated from the castor bean (*Ricinus communis*) is one of the most potent and lethal molecules known. Castor beans are processed worldwide on an industrial scale for the castor oil. Ricin, a byproduct of castor oil, is a real threat for bioterrorism and for biological warfare, especially when dispersed by aerosol. There are no FDA approved vaccines or therapeutics to protect against ricin or the related Shiga toxins, which cause food poisoning and dysentery in millions of people around the world. Ricin is a type II ribosome inactivating protein (RIP), which consists of an active A chain (RTA) covalently linked to a cell binding B chain (RTB). RTA inhibits protein synthesis by removing a specific adenine from the highly conserved α-sarcin/ricin loop (SRL) in the large rRNA and inhibits protein synthesis. RTA-antibody complexes have been explored as immunotoxins against cancer cells. A thorough understanding of how ricin enters cells and traffics to the ribosome, how it inactivates ribosomes with near perfect efficiency, how it induces inflammatory signaling pathways, and programmed cell death is critical for understanding the complexity of ricin and for reducing its toxicity. The eight articles published in this issue address these research needs and provide important insights into the mechanisms of the toxicity of ricin. They will contribute to the design of therapies against intoxication by ricin and related toxins.

Polito et al. [1] review the history of ricin starting from its use in traditional and folk medicine and highlight the research milestones in the characterization of enzymatic activity, structure, toxicity, and medical applications [1]. Ricin is rapidly internalized and catalytic amounts are needed to inhibit protein synthesis. It has been used as a powerful tool to understand intracellular trafficking and cell death pathways. Sowa-Rogozinska et al. [2] review the current knowledge about the intracellular transport of ricin and identification of host factors that facilitate transport to increase our understanding of the mechanism of the cytotoxicity of ricin. This review summarizes medical applications of ricin and highlights its role as a valuable component of immunotoxins against cancer [2].

Previous studies identified the host target of ricin as the ribosomal P stalk [3,4] and showed that binding to the P stalk is necessary for depurination of the SRL by RTA on intact ribosomes [5]. The eukaryotic P stalk contains P0 protein and two P1–P2 dimers with identical C-terminal sequences, which are critical for interaction with the translation factors and factor dependent GTP hydrolysis. Ricin binds to the C-termini of the human P1–P2 dimer, which represents the smallest component of the eukaryotic stalk [6]. Grela et al. [7] present the current understanding of the structure and function of the ribosomal stalk and the consequence of ricin dependent depurination of the SRL on ribosome performance and translation.

Small molecules that can enter and rescue intoxicated cells by inactivating intracellular ricin are highly sought after as countermeasures. Although small-molecule RIP inhibitors have been identified, none of them exhibited potent protection against RIPs. Li et al. addressed if peptides mimicking the conserved C-terminal sequences of P proteins will inhibit the activity of RTA by preventing its interaction with the ribosome [8]. They show that these peptides interact with the ribosome binding site of RTA and inhibit the activity of RTA by disrupting its interaction with ribosomes [8]. These results establish the ribosome binding site of RTA as a new target for inhibitor discovery [8].

Ricin inhalation causes acute lung injury characterized by a massive inflammatory response. Hodges et al. [9] evaluated the cell death modulatory activity of cytokines in ricin toxicity in human lung epithelial cells and showed that tumor necrosis factor (TNF) family cytokines induce distinct cell death pathways when administered in combination with ricin [9]. Targeting these cell death pathways may lead to novel therapeutic approaches to ricin toxicity [9]. The use of neutralizing antibodies is a promising post-exposure treatment against ricin intoxication. Falach et al. [10] generated equine derived antibodies against ricin for post exposure treatment. They generated an inactivated toxin and constructed monomerized ricin antigen by irreversible reduction of the A and B subunits. Immunization of a horse with the monomerized toxin yielded high titers of neutralizing antibodies. Passive immunization of mice with equine derived F(ab')$_2$ based antitoxin conferred protection against a lethal intranasal ricin challenge [10].

Ricin is a therapeutic agent and a potential threat to public health and safety. Several methods to detect ricin have been developed; however, each method has its limitations [11]. Innovative assays for toxin detection and mitigation are needed. Micro RNA (miRNA) profiles can help understand ricin toxicity mechanisms and could serve as potential biomarkers for ricin intoxication. Pillar et al. [12] investigate the effect of pulmonary exposure of mice to ricin on miRNA expression profiles in mouse lungs. They show significant changes in the lung tissue expression levels of miRNAs involved in innate immunity pathways. They confirm these findings by gene expression analysis and show activation of immune regulation pathways and immune cell recruitment after ricin exposure [12]. Sousa et al. [13] describe an accelerated solvent extraction (ASE) method followed by matrix-assisted laser desorption/ionization time-of-flight mass spectrometry (MALDI-TOF-MS) and MALDI-TOF-MS/MS for extraction and detection of ricin in forensic samples. This method could also detect ricin in gamma-irradiated samples [13].

The papers in this issue provide readers with a better understanding of ricin trafficking, ribosome binding, SRL depurination, cell signaling, toxicity, and ricin detection mechanisms and identify new targets that may be useful in the development of ricin antidotes.

Conflicts of Interest: The author declares no conflict of interest.

References

1. Polito, L.; Bortolotti, M.; Battelli, M.G.; Calafato, G.; Bolognesi, A. Ricin: An Ancient Story for a Timeless Plant Toxin. *Toxins* **2019**, *11*, 324. [CrossRef] [PubMed]
2. Sowa-Rogozinska, N.; Sominka, H.; Nowakowska-Golacka, J.; Sandvig, K.; Slominska-Wojewodzka, M. Intracellular Transport and Cytotoxicity of the Protein Toxin Ricin. *Toxins* **2019**, *11*, 350. [CrossRef] [PubMed]
3. Chiou, J.C.; Li, X.P.; Remacha, M.; Ballesta, J.P.; Tumer, N.E. The ribosomal stalk is required for ribosome binding, depurination of the rRNA and cytotoxicity of ricin A chain in Saccharomyces cerevisiae. *Mol. Microbiol.* **2008**, *70*, 1441–1452. [CrossRef] [PubMed]
4. May, K.L.; Li, X.P.; Martinez-Azorin, F.; Ballesta, J.P.; Grela, P.; Tchorzewski, M.; Tumer, N.E. The P1/P2 proteins of the human ribosomal stalk are required for ribosome binding and depurination by ricin in human cells. *FEBS J.* **2012**, *279*, 3925–3936. [CrossRef] [PubMed]
5. Li, X.P.; Kahn, P.C.; Kahn, J.N.; Grela, P.; Tumer, N.E. Arginine residues on the opposite side of the active site stimulate the catalysis of ribosome depurination by ricin A chain by interacting with the P-protein stalk. *J. Biol. Chem.* **2013**, *288*, 30270–30284. [CrossRef] [PubMed]
6. Grela, P.; Li, X.P.; Horbowicz, P.; Dzwierzynska, M.; Tchorzewski, M.; Tumer, N.E. Human ribosomal P1-P2 heterodimer represents an optimal docking site for ricin A chain with a prominent role for P1 C-terminus. *Sci. Rep.* **2017**, *7*, 5608. [CrossRef] [PubMed]
7. Grela, P.; Szajwaj, M.; Horbowicz-Drozdzal, P.; Tchorzewski, M. How Ricin Damages the Ribosome. *Toxins* **2019**, *11*, 241. [CrossRef] [PubMed]
8. Li, X.P.; Kahn, J.N.; Tumer, N.E. Peptide Mimics of the Ribosomal P Stalk Inhibit the Activity of Ricin A Chain by Preventing Ribosome Binding. *Toxins* **2018**, *10*, 371. [CrossRef] [PubMed]

9. Hodges, A.L.; Kempen, C.G.; McCaig, W.D.; Parker, C.A.; Mantis, N.J.; LaRocca, T.J. TNF Family Cytokines Induce Distinct Cell Death Modalities in the A549 Human Lung Epithelial Cell Line when Administered in Combination with Ricin Toxin. *Toxins* **2019**, *11*, 450. [CrossRef] [PubMed]

10. Falach, R.; Sapoznikov, A.; Alcalay, R.; Aftalion, M.; Ehrlich, S.; Makovitzki, A.; Agami, A.; Mimran, A.; Rosner, A.; Sabo, T.; et al. Generation of Highly Efficient Equine-Derived Antibodies for Post-Exposure Treatment of Ricin Intoxications by Vaccination with Monomerized Ricin. *Toxins* **2018**, *10*, 466. [CrossRef] [PubMed]

11. Zhou, Y.; Li, X.P.; Kahn, J.N.; Tumer, N.E. Functional Assays for Measuring the Catalytic Activity of Ribosome Inactivating Proteins. *Toxins* **2018**, *10*, 240. [CrossRef] [PubMed]

12. Pillar, N.; Haguel, D.; Grad, M.; Shapira, G.; Yoffe, L.; Shomron, N. Characterization of MicroRNA and Gene Expression Profiles Following Ricin Intoxication. *Toxins* **2019**, *11*, 250. [CrossRef] [PubMed]

13. Sousa, R.B.; Lima, K.S.C.; Santos, C.G.M.; Franca, T.C.C.; Nepovimova, E.; Kuca, K.; Dornelas, M.R.; Lima, A.L.S. A New Method for Extraction and Analysis of Ricin Samples through MALDI-TOF-MS/MS. *Toxins* **2019**, *11*, 201. [CrossRef] [PubMed]

 *toxins*

Review

Ricin: An Ancient Story for a Timeless Plant Toxin

Letizia Polito [†,*], Massimo Bortolotti [†], Maria Giulia Battelli [†], Giulia Calafato and
Andrea Bolognesi *

Department of Experimental, Diagnostic and Specialty Medicine—DIMES, General Pathology Section,
Alma Mater Studiorum—University of Bologna, Via S. Giacomo 14, 40126 Bologna, Italy;
massimo.bortolotti2@unibo.it (M.B.); mariagiulia.battelli@unibo.it (M.G.B.); giulia.calafato2@unibo.it (G.C.)
* Correspondence: letizia.polito@unibo.it (L.P.); andrea.bolognesi@unibo.it (A.B.);
   Tel.: +39-051-209-4700 (L.P. & A.B.)
† These authors contribute equally to this work.

Received: 15 April 2019; Accepted: 5 June 2019; Published: 6 June 2019

Abstract: The castor plant (*Ricinus communis* L.) has been known since time immemorial in traditional
medicine in the pharmacopeia of Mediterranean and eastern ancient cultures. Moreover, it is still
used in folk medicine worldwide. Castor bean has been mainly recommended as anti-inflammatory,
anthelmintic, anti-bacterial, laxative, abortifacient, for wounds, ulcers, and many other indications.
Many cases of human intoxication occurred accidentally or voluntarily with the ingestion of castor
seeds or derivatives. Ricinus toxicity depends on several molecules, among them the most important
is ricin, a protein belonging to the family of ribosome-inactivating proteins. Ricin is the most studied
of this category of proteins and it is also known to the general public, having been used for several
biocrimes. This manuscript intends to give the reader an overview of ricin, focusing on the historical
path to the current knowledge on this protein. The main steps of ricin research are here reported,
with particular regard to its enzymatic activity, structure, and cytotoxicity. Moreover, we discuss ricin
toxicity for animals and humans, as well as the relation between bioterrorism and ricin and its impact
on environmental toxicity. Ricin has also been used to develop immunotoxins for the elimination
of unwanted cells, mainly cancer cells; some of these immunoconjugates gave promising results in
clinical trials but also showed critical limitation.

Keywords: bioterrorism; cancer therapy; castor bean; folk medicine; immunotoxins; plant toxins;
ribosome-inactivating proteins; ricin; rRNA N-glycosylase activity; traditional medicine

Key Contribution: Despite the large number of papers on ricin, it is often difficult for inexpert readers
to have a global view on this toxin. This manuscript intends to give an overview of ricin. Starting
from the use of Ricinus plant in traditional and folk medicine, we highlight the milestones of research
on ricin, with particular regard to its enzymatic activity, structure, cytotoxicity, toxicity for animals
and humans and the double face of its employ, for biocrimes and medicine.

---

1. Castor Bean in Traditional and Folk Medicine

Ricin derives from *Ricinus communis* L. (Euphorbiaceae family), also known as castor bean or
palma Christi. The genus Ricinus has only one known species: the castor oil plant. The plant possibly
originates from Africa and Asia and now is widespread throughout temperate, subtropical, and tropical
areas, growing as an invasive plant or being cultivated for different purposes.

The castor plant has been known since time immemorial and its use in the prehistoric era has been
evidenced by archaeological findings such as that of the Border Cave in South Africa. Traces of wax
containing ricinoleic and ricinelaidic acids were found on a thin wooden stick, which was suggested to
be a poison applicator, dating back to about 24,000 years ago [1]. The castor seeds and other parts of

the castor plant were certainly utilized in ancient Egypt for pharmacological purposes. In the Ebers Papyrus, an Egyptian medical treatise dating back to before 1500 BCE, an entire chapter is dedicated to the castor bean that is indicated as an abortifacient, a laxative, a remedy for abscessual illness, baldness, and so on [2]. In the Hearst Papyrus, written approximately in the same period, various castor plant parts are included as ingredients in some prescriptions for internal use, with the aim of expelling fluid accumulation or promoting diuresis, as well as for external use as poultices for bandaging [3]. Ancient Egyptians knew the toxicity of castor bean and the use of seed pulp, included in drug preparations for oral ingestion, was recommended only in small amounts. In addition, a castor seed-containing concoction was prescribed to cure the urinary disease of a possibly diabetic child [4]. Around 400 BCE, the father of western medicine Hippocrates prescribed castor bean oil for laxative and detoxifying action [5]. The Greek herbalist and physician Pedanius Dioscorides (40 to 90 CE) in *De Materia Medica* wrote that castor seeds could be used as expectorant, diuretic, emetic, laxative, anti-inflammatory, to cure erysipelas, burns, varicose veins, etc. [6]. In the same period, Pliny the Elder (23 to 79 CE) wrote *Naturalis historia*, comprising the whole area of antique knowledge. In this encyclopedic work, also castor bean found a place [7].

Castor bean was used also in the pharmacopeia of eastern ancient cultures. In Chinese traditional medicine, castor seeds were recommended for their anthelmintic activity; seed poultice and leaf juice were prescribed for external use to treat ulcers and chronic wounds, whereas the latex was instilled in the ear for rhinitis treatment (reviewed in [8]). In Ayurveda, castor plant is used for rheumatic conditions, as well as for gastropathy, constipation, inflammation, fever, ascites, bronchitis, cough, skin diseases, colic, and lumbago. In Yunani medicine, castor root is used as a purgative and for skin diseases, the leaves are used to increase breastmilk production and are applied to skin for burns, the seeds and the oil act as a purgative, useful in liver troubles, pains, lumbago, boils, piles, ringworm, inflammation, ascites, asthma, rheumatism, dropsy, and amenorrhea (reviewed in [9]). Ground castor seeds or leaf paste have been applied in veterinary medicine to heal sprains, swelling, and wounds [10].

Castor bean has been used in folk medicine throughout the world and has been reported: (i) As a galactogogue on the Mediterranean coasts of Europe, where fresh leaves or leaf juice are applied on the puerperal breast to promote lactation; (ii) as a remedy for various articular, cutaneous, or ocular diseases in Africa, where crushed seeds or oil, sometimes in combination with other plants, are spread or rubbed on the part of the body in need, or a root decoction is drunk to induce uterine contraction as an abortive; (iii) as a medicament to cure erysipelas, flu, inflammation of the womb, and stomach aches in the Caribbean, where a leaf poultice is recommended; (iv) as an anthelmintic or a purgative in Brazil where the seed oil is orally consumed, or locally applied with the purpose of stopping hair loss, healing wounds, or burns (reviewed in [11]).

The laxative and abortifacient activities of castor seeds have been attributed to the activation of intestinal and uterine smooth-muscle cells via prostaglandin EP3 receptors induced by ricinoleic acid [12]. Castor oil-induced diarrhea can be antagonized by hexane extract of *Citrus limon* peel that activates antisecretory and antimotility mechanisms through the β adrenergic system [13]. The purgative and anthelmintic actions of the oral ingestion of castor seeds, at least in part, have been ascribed to the irritating effect caused to the intestine by ricin, as reported in toxicological studies (reviewed in [14]). In addition, the antiflogistic action of castor bean could be related to the high toxicity of ricin to macrophagic cells, which are responsible for producing inflammatory cytokines (reviewed in [15]). This effect, together with the anti-pathogen activity of ricin, could promote healing of the lesions, thus justifying its use in the treatment of various skin conditions.

2. The Ricin Story

Castor seed toxicity began to be investigated at the end of nineteenth century at Schmiedeberg's laboratory in Strasbourg. The toxic component of Ricinus could be extracted with water and precipitated with alcohol, but it lost its toxic activity through heating, treatment with strong acid, or repeated precipitation with alcohol. In 1887, Dixson supposed that the toxicity of Ricinus was due to an albumen-like toxic body [16]. However, it was still unknown whether the seed toxicity was due to a protein or a glycoside (reviewed in [17]). The problem was solved at the Medical Faculty of Dorpat (now Tartu) where an extremely toxic protein was partially purified from castor seed or press cake and named ricin. This finding was published in the doctoral thesis written by Hermann Stillmark under the supervision of Prof. Rudolf Kobert [18]. Stillmark noticed the agglutinating activity of ricin on red blood cells, that had mistakenly been believed to be the cause of ricin toxicity until the agglutinin was separated from the toxin [19].

Paul Ehrlich began his experiments in immunology by feeding mice with small amounts of ricin or abrin, another similar plant toxin, until they were accustomed and became resistant to the toxin used, yet still remaining sensitive to the other toxin. The immunization was strictly specific, started after a few days, and persisted at least for several months [20,21]. He was successful in the production of antisera against abrin and ricin and in the determination of antibody titer in serum and milk. Ehrlich drew animal experiments that clarified the transmission of passive immunity from mother to offspring through the transplacental transfer of antibodies and through breastfeeding. He investigated the dynamics of the antibody response and was the first to envisage the presence of binding sites on the cell surface (reviewed in [22]). These studies, together with those on the immunity to bacterial toxins, led him to formulate his side-chain theory of antibody formation and to win, in 1908, the Nobel Prize [23].

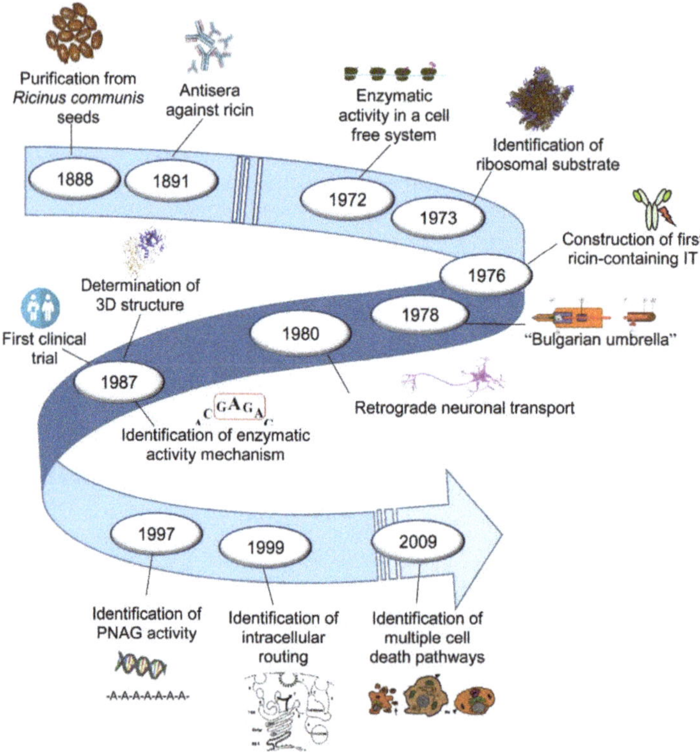

Figure 1. The main milestones of ricin research.

Interest in ricin was rekindled when the anticancer activity of this toxin on Ehrlich ascites cells in a mouse model was published [24]. A strong inhibition of protein synthesis by ricin was observed in cultures of both Ehrlich ascites tumour cells and Yoshida ascites hepatoma cells. The inhibition of protein synthesis by ricin requires more time in rat liver than in neoplastic cells [25]. The prospect of a possible use in cancer therapy highlighted the need to investigate which part of the proteosynthetic machinery was damaged and how the toxin managed to enter the cell to reach its target. Hereinafter, we highlight the milestones of research on ricin, with particular regard to its enzymatic activity, structure, cytotoxicity, toxicity for animals and humans, and its use as an immunotoxin, used in experimental models and in clinical trials. The main milestones are shown in Figure 1.

2.1. Ricin Structure

The first information about the bi-chain nature of ricin structure dates to the early 1970s, when it was shown that ricin was composed of two chains, A (active) and B (binding), linked together through a disulphide bond [26,27]. In the same period, the complete primary sequence of the ricin A and B chains was determined [28,29]. Ricin holotoxin structure was solved for the first time at 2.8 Å resolution (Figure 1) [30]. This pioneering work demonstrated that ricin A chain was a globular protein folded into three domains all contributing to the active site, while the B chain lectin folded into two domains, each binding lactose in a shallow cleft. The interface between the A and B chains showed some hydrophobic contact in which proline and phenylalanine side chains played a prominent role. Four years later, the same researchers refined ricin structure at 2.5 Å (Figure 2a), allowing a more detailed molecular description of the holotoxin and of the separated A and B chains [31–33]. Ricin A chain has been described as a globular protein consisting of 267 amino acids and organized in 8 α-helices and 8 β-strand structures. Ricin B chain consists of 262 amino acids and two homologues domains, each containing a lactose binding site and several areas of amino acid homology, possibly derived from a gene duplication. In 1995, after purification of a complex of ricin A chain cross-linked to the ribosome, it was found the binding of ricin A chain with the ribosomal proteins L9 and L10e [34,35].

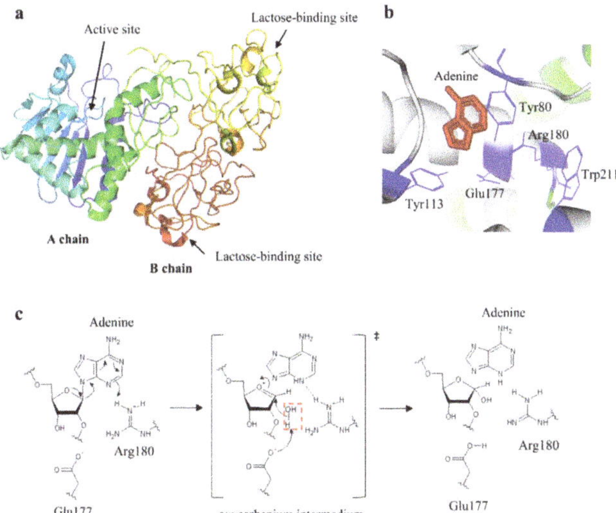

Figure 2. (**a**) Ribbon model of the crystal structure of ricin at 2.5 Å (accession number Protein Data Bank 2AAI). The A chain domains are colored in green, blue, and light blue; the B chain domains are colored in yellow and orange. (**b**) Catalytic site of ricin. The key residues are indicated and colored in blue, whereas adenine substrate is depicted in red. (**c**) Proposed mechanism of depurination reaction catalyzed by ricin. The hydrolysis proceeds through a dissociative mechanism forming an oxocarbenium transition state. Arg180 protonates the leaving group and the N-glycosidic bond is broken. Glu177 deprotonates the hydrolytic water (highlighted by a red dotted rectangle) that attacks carbon to complete the depurination reaction. Figure 2a and 2b were produced by PyMOL (version 2.3.1); Figure 2c was produced by ACD/ChemSketch (version 2015.2.5).

The knowledge of the tridimensional structure of ricin yielded more information on its active site. Studies based on the formation of complexes between the A chain, both native and recombinant, and adenine-containing nucleotides allowed for the identification of key residues in enzymatic activity. In particular, Tyr80, Tyr123, Glu177, Arg180, and Trp211 were found to form the binding site for adenine (Figure 2b) [30,36]. In the 1990s, the molecular mechanism of adenine release was hypothesized: Adenine is sandwiched between Tyr80 and Tyr123 in a π stacking interaction; the N3 of adenine is protonated by Arg180, promoting the C1'-N9 bond breaking, and thus forming an oxocarbenium moiety on the ribose (Figure 2c) [36,37]. This transition state is stabilized by Glu177; a water molecule lies on the opposite side of the sugar ring from adenosine, which will be polarized by Arg180 to a hydroxide character that rapidly attacks the sugar carbon completing the reaction.

2.2. Ricin Enzymatic Activity

The introduction of a cell-free system utilizing a lysate from rabbit reticulocytes [38] helped to clarify that ricin inhibited the peptide chain elongation (Figure 1) [27]. The two polypeptides showed different properties: The A chain possessed the toxic activity, while the B chain was a galactose-specific lectin binding the cell surface [26]. Treating the toxin with reducing agents resulted in more activity in inhibiting cell-free protein synthesis [39]. Firstly, the target of the toxic action was identified as the ribosome (Figure 1), then as the 60 S subunit of eukaryotic ribosome [40], which became unreactive toward elongation factors [41]. The toxin was found to prevent the binding between elongation factors and ribosomes avoiding the subsequent elongation-factor-dependent GTPase activity [41,42]. The A-chain molecule was very active on its substrate and it was calculated that one molecule can inactivate 2000 ribosomes/min, with a K_m of 0.1–0.2 mM [43].

In addition to ricin, several other plant proteins have been identified to possess a similar protein synthesis inhibiting action. Most of them had a single polypeptide chain similar to the A chain of ricin. They were called Ribosome-Inactivating Proteins (RIPs) (reviewed in [44,45]).

The already supposed enzymatic nature of ricin A chain was finally demonstrated in 1987 by Endo et al. who discovered that ricin A-chain cleaved the N-glycosidic bond of an adenine residue, A4324 in rat 28 S RNA, from the ribose of a highly conserved ribosomal RNA single-stranded loop involved in the binding of elongation factors (Figure 1). The toxin did not directly break the RNA chain, but the depurinated RNA was susceptible to hydrolysis [46,47]. Consequently, ricin activity was identified as an rRNA N-glycosidase (EC 3.2.2.22).

Following this, it was demonstrated that the enzymatic activity of RIPs was broader than previously described. All tested RIPs were able to release adenine from DNA, in addition to rRNA, and some of them were also able to act on other polynucleotide substrates, releasing adenine from the sugar phosphate backbone of polynucleotide substrates (Figure 1) [48,49]. For this reason, the name of adenine polynucleotide glycosylase was proposed for RIPs. Thus, the ability of acting on various substrates and extensively depurinating some of them, suggested that the protein synthesis inhibition could be only one of the ways of RIP-mediated cell killing. Ricin was shown to be able to release adenine from rRNA, DNA (chromatin and naked), and also poly(ADP-ribosyl)ated poly(ADP-ribose) polymerase, an enzyme involved in DNA repair [48,50]. Furthermore, it was observed that many RIPs were able to cleave more than one adenine: ricin was able to detach few adenines from the DNA (tens), but some single-chain RIPs were able to detach even thousands of them. The hypothesis that ricin could act directly on DNA in cellular models was strengthened by the evidence that damage to nuclear DNA, consistent with the enzymatic activity (adenine release) on DNA in cell-free systems, was concomitant with protein synthesis inhibition and preceded apoptosis [51].

2.3. Ricin Cellular Uptake, Routing, and Toxicity

Starting from the mid-1970s, several research groups focused on ricin binding and internalization studies, demonstrating that the interaction of ricin with the cell started from the binding of the B chain to galactosyl residues on the cell surface, allowing access to the endosomal compartment [52]. Ricin binds to both glycolipids and glycoproteins with terminal galactose. Since ricin binds to a variety of different molecules, it seems to be internalized by different endocytic pathways, as well as by using different pathways to reach the Golgi apparatus to intoxicate the cell. In HeLa cells, about 10^7 binding sites were found for ricin, but only small amount of the bound toxin reached the Golgi network and participated in cell intoxication [52].

Firstly, it was reported that ricin entered into cytoplasm through clathrin-dependent endocytosis [53]. Afterwards, it became clear that clathrin-independent mechanisms were also involved [54]. After cell uptake, ricin is delivered to early endosomes, from where most of protein molecules are recycled back to the cell surface or delivered, via late endosomes, to lysosomes for proteolytical degradation. A small amount of non-degraded ricin is addressed within the trans-Golgi network [55]. The involvement of the Golgi complex in ricin routing was confirmed using different Golgi-disrupting agents, such as brefeldin A, monensin, etc. In fact, the pretreatment with these agents inhibited the cytotoxic effects of ricin [56]. It was demonstrated that ricin was cycled from Golgi to the endoplasmic reticulum via coatomer protein 1 (COP-1)-coated vesicles [57], although it was later proved that the COP-1-independent pathway could also be involved [58].

The complete elucidation of intracellular ricin traffic occurred when it was demonstrated that, after reaching the endoplasmic reticulum, the two ricin chains were separated, and the A chain was retro-translocated through the quality control pathway delivering misfolded proteins to cytosol (Figure 1) [59]. Recently, it has been demonstrated that cholesterol rafts are required for Golgi transport of ricin; meaning that glycosphingolipids may not be required (reviewed in [60]).

The portion of A chain that quickly refolded, thus avoiding ubiquitination and proteosomal degradation, was able to reach its intracellular target (reviewed in [61]). It was estimated that one molecule of active ricin that arrives to its substrate is enough to kill one cell [62].

The discovery that ricin, and some related toxins, may be retrogradely transported along neuronal processes (Figure 1) [63] opened a new field of research in neurobiology and this property has been exploited for the selective destruction of neuron bodies.

Different cell types have shown variable levels of sensitivity to ricin (reviewed in [14]), possibly because of the mannose receptor expression on the cell surface and endocytosis efficacy. Ricin has been shown to be one of the most toxic plant toxins on cell lines with $IC_{50}s$ (concentration inhibiting protein synthesis by 50%) ranging from less than 0.1 to 1 pM [26,64–66]. However, it must be taken into account that it is very difficult to make a direct comparison of the data available in the literature about ricin cytotoxicity, because of the differences in the experimental approaches and technical conditions.

The polynucleotide depurinating activity of RIPs suggests the possibility of a wider toxic action on many biological substrates, not excluding the induction of oxidative stress. This could explain the induction of more than one cell death pathway, e.g. apoptosis and necroptosis, caused by ricin and other RIPs (Figure 1) [64,67].

3. Ricin Toxicity in Humans and Animals

On one hand, ricin has been studied for bio-medical applications, exploiting the ability of the A-chain to kill target cells once linked to a monoclonal antibody, as below described in the immunotoxins chapter. On the other hand, ricin has attracted nefarious interests, with a history of military, criminal, and terroristic uses [68].

The acute toxicity of ricin is highly variable depending on the animal species and strain. The pathological effects and subsequent clinical signs of ricin intoxication depend also on the route of exposure, as this dictates the subsequent tissue distribution of the toxin. Following intravenous or intramuscular administration, lesions eventually develop in the spleen, liver, and kidney whilst the lung remains unaffected. After oral ingestion, the gastrointestinal tract is severely affected. Inhalational exposure produces effects that are mainly confined to the respiratory tract [69].

The majority of data on animal toxicity has been derived from laboratory experiments in rodents, principally rat and mouse models. Oral administration of ricin was reported to give a lethal dose (LD) for 50% of animals (LD$_{50}$s) 20 to 30 mg/kg in rat and 15 to 35 mg/kg in mouse [70–72]. For intravenous, inhalation and intraperitoneal routes, toxicity is approximately 1000-fold higher than that obtained for the oral route, with LD$_{50}$ values in mouse of 2 to 10 µg/kg, 3–5 µg/kg and 22 µg/kg, respectively [70,73]. The lower toxicity of ricin after oral exposure is due to the protein destruction in the lumen of the intestinal tract [74,75]. Ricin acts in a time- and concentration-dependent manner. Notably, there is a time delay of about 10 h before death occurs, even when very high doses are applied [76].

3.1. Oral Toxicity

In humans, most intoxications occurred accidentally or voluntarily with the ingestion of castor seeds; only a few cases of intentional absorption of castor bean extracts have been documented in suicide attempts [76]. Whole-ingested beans can pass intact through the gastrointestinal tract, whereas chewing facilitates ricin release. Also, it has been reported that the seed can act as 'timed-release' capsule for the toxin, allowing its release in the lower bowel, where it causes more damage [72]. After ingestion, vomiting, diarrhea, and abdominal pain are common symptoms. Massive gastrointestinal fluid and electrolyte loss are described, often complicated by hematemesis or melaena. Finally, hypovolemic shock and multiorgan failure occur, which particularly involves the spleen, liver, and kidney [77,78].

Despite the high number of intoxicated subjects with castor beans, it is quite difficult to calculate LD values for ricin in humans. In fact, the effective ingested ricin dose can only be supposed, because of ricin content variations depending on the size, weight, and moisture of the seeds, as well as on cultivar, region, season, and plant growth stage. Moreover, in intoxicated subjects, it must be taken into account the degree of mastication, stomach content, age, and comorbidities, parameters that are obviously more heterogeneous compared to experimental poisoning of animals. Considering all these parameters, the fatal oral dose of ricin in humans has been estimated to range from 1 to 20 mg/kg (approximately 5 to 10 beans) [70,79].

3.2. Inhalation Toxicity

No data are available for human ricin uptake by inhalation. In non-human primates, LD$_{50}$ has been estimated to be 5 to 15 µg/kg depending on aerosol particle size. Inhalation of particles that are able to penetrate deeply into the lungs (1 to 5 µm diameter) display much more toxicity than larger particles [72,80]. Inhalation of ricin causes slow onset of respiratory distress (difficulty breathing), coughing, fever, pulmonary lesions, and edema. Intoxicated animals develop fibrinopurulent necrotizing pneumonia accompanied by necrotizing lymphadenitis, typically after a dose-dependent delay of 8 to 24 h. Death occurs as a result of respiratory failure due to massive alveolar fluid accumulation. The liver, kidney, and small intestines appear congested, although little histologic changes have been shown [72,80,81].

3.3. Parenteral Toxicity

Data regarding parenteral ricin intoxication derive mainly from animal studies. By injection, mice had an LD$_{50}$ of 3 to 5 µg/kg by intravenous and 22 µg/kg by subcutaneous route [82], rabbits had LD$_{50}$ 0.5 µg/kg by the intravenous route and 0.1 µg/kg by the intramuscular route, while guinea pigs had LD$_{50}$ <1.1 µg/kg by the intravenous route and 0.8 µg/kg by the intramuscular route [83]. Human data only derives from the few cases of suicide or murder, or their attempt; the most known episode is the assassination of the Bulgarian dissident Georgi Markov, who in 1978 died three days after possibly being stabbed with an umbrella loaded with a ricin-containing pellet (Figure 1) [84].

By parenteral administration, immediate local pain at the injection site is reported, followed by general weakness within five hours. The following symptoms, that are general and maybe similar to sepsis (fever, headache, dizziness, anorexia, nausea, vomiting, hypotension, abdominal pain), can be delayed for as much as 10 to 12 h, even with high doses. Usually local tissue damage at the site of the injection is observed. Laboratory abnormalities included elevated liver transaminases, amylase and creatinine kinase, hyperbilirubinemia, myoglobinuria, and renal impairment. The clinical course may progress to multisystem organ failure. Preterminal complications included gastrointestinal hemorrhage, hypovolemic shock, and renal insufficiency [78,84].

4. Bioterrorism and Environmental Toxicity

Ricin is currently monitored as Schedule 1A of the Chemical Weapons Convention (CWC) and is a Category B substance under the Biological and Toxins Weapons Convention (BTWC) [80]. Despite its toxicity, ricin is less potent than other agents, such as botulinum neurotoxin or anthrax. It has been estimated that eight tons of ricin would have to be aerosolized over a 100 km^2 area to achieve about 50% casualty, whereas only a kilogram of anthrax spores would cause the same effect [85]. Thus, deploying an agent such as ricin over a wide area, although possible, becomes impractical from a logistics standpoint. However, the availability of castor beans and the quite simple procedure for rough ricin purification have attracted criminal and terrorist interest for small scale biocrimes or to cause collective media-driven alarm [80].

From castor seeds, a nontoxic oil can be extracted that has multitude of uses in many sectors, including cosmetic, pharmaceutic, mechanical, and chemical industry. Castor oil production is increasing worldwide because of its versatile application, low cost, availability, and biodegradability. In addition, the oil-free seed pulp can be used in agriculture as a natural fertilizer [86], although the processing of castor seeds requires great caution due to the high allergenicity [87,88] and extreme toxicity [76] of their protein fraction, represented, above all, by ricin. World production of castor oil increased from 0.8 million tons in 2000 [89] to 1.21 million tons in 2014 [90], with a castor seed production of 1.49 million tons in 2017 [91]. Leading producing countries are India, with over 80% of the global yield, Mozambique, China, Brazil, Myanmar, Ethiopia, Paraguay, and Vietnam [92]. The oil makes up about 50% of the weight of the seeds and is mostly constituted of ricinoleic acid (90%), with minor amounts of dihydroxystearic, linoleic, oleic, and stearic acids. Ricin isoforms and the alkaloid ricinine, are not transferred to the oil fraction during extraction, which can be performed by cold or warm pressing, but remain in the seed cake [93,94].

Castor bean meal press cake or other residues of the castor oil production have been employed as a protein source for feed or fertilizer, but their use is very limited due to ricin toxicity [76]. In 2008, the European Food Safety Agency defined ricin as an undesirable substance in animal feed. Ricinus derived material should be appropriately inactivated through physical and/or chemical methods to guarantee animal and human health [95]. Nevertheless, many accidental poisonings are still reported for animals eating improperly detoxified fertilizer or other agricultural products containing castor derived material [76,94].

In order to block the toxic action of ricin, different strategies have been evaluated: Vaccines, inhibitors, and passive immunity. Vaccines against ricin with the consequent production of neutralizing antibodies did not give satisfactory results in vivo (reviewed in [96]). Inhibitors of ricin can block the active site or work as a substrate analogue; however, the available data are limited to in vitro experiments [97]. More recently, inhibitors of cell routing have been developed, sometimes giving promising results, also in vivo [98,99]. To date, passive immunity has proven to be the only effective strategy for treating intoxication caused by ricin. The delay in the appearance of signs of intoxication makes confirmation of exposure, diagnosis of intoxication, and the subsequent medical response technically and logistically challenging. The development of anti-ricin sera or antibodies, effective even when used several hours after toxin exposure, represents a step forward in treatment of ricin intoxication, as it increases the time window of intervention (WOO, window of opportunity). Many authors described effective post-exposure treatment of ricin intoxication with specific antibodies, but with a limited WOO (~8 h) [100–103]. Other authors reported a survival between 50% and 89% of mice treated with anti-sera 24 h after intoxication [104–106]. Once internalized into the cells, ricin cannot be neutralized by antibodies, thus limiting the therapeutic window. However, Whitfield et al. in 2017 reported 100% protection in aerosolized ricin-treated mice with a single administration of a F(ab')$_2$

polyclonal ovine antitoxin given 24 h post-exposure [69]. Even when performed in the same animal species, comparison between diverse experiments is often difficult, due to the different toxin dose and route of administration utilized. Moreover, there are few data about correlation between the antitoxin dose required for protection and the WOO.

5. Ricin-Containing Immunotoxins

Many researchers have tried to exploit the high cytotoxicity of ricin for medical purposes to eliminate pathological cells. Although ricin possesses highly efficient cell killing mechanisms, it lacks selectivity towards cell targets. In order to increase selectivity, the possibility of linking ricin to carriers specific for targets on unwanted cells has been explored. The most widely used carriers are antibodies and the corresponding conjugates are referred to as immunotoxins (ITs).

The first IT, created in 1976 by Moolten and co-workers, was made by Ricin Toxin-A chain (RTA) linked to a rat tumor-specific antibody against a rat lymphoma, namely (C58NT)D (Figure 1) [107]. To date, a multitude of pre-clinical and clinical studies have shown the potential use of several ricin-ITs towards different cancer types, from hematological to solid ones, and towards normal cells, unwanted due to them being responsible for a pathological state (reviewed in [108,109]). Different approaches have been used, over time, to generate ITs. In the first strategy, ITs were composed by the antibody chemically linked to the entire RIP and they were used for in vitro studies showing high cytotoxicity [110]. Despite the high in vitro efficiency, the relevant non-specific toxicity reported in vivo, due to the characteristics of the lectin chain, brought researchers to sterically block, chemically modify, or remove the B chain, thus balancing toxicity and specificity. In 1985, Weil-Hillman et al. tested an anti-Mr 67,000 protein linked to either blocked-chain B ricin or RTA, reporting interesting results in vitro, but not in vivo in a nude mouse model [111]. The 1980s were years of great ferment for molecular biology and genetic engineering, paving the way for the second generation of ITs. Many researchers tried to improve the IT penetration in tumor mass by reducing the antibody size, using antigen-binding (Fab), or variable (Fv) fragments instead of entire antibodies. In 1988, Ghetie et al. created a new IT composed by Fab conjugated to chemically deglycosylated RTA (dgA) [112]. A few years later, they used an anti-CD122-dgA IT in SCID-Daudi mice, showing promising results since the IT was able to specifically kill tumor cells in vivo, extending the mean survival time up to 57.9 +/− 3.8 days [113]. Moreover, FitzGerald et al. described the antitumor activity of recombinant RTA (rRTA) linked to anti-mouse transferrin receptor in a nude mouse model of human ovarian cancer. Animals treated with IT had an extended life span from 45 (lower doses) to 70/80 days (higher doses) [114]. Finally, in 1997 the first ricin-containing recombinant immunotoxin (rIT) was obtained through the expression of a fusion gene composed by sequences encoding anti-CD19-FVS191 (single-chain Fv), cathepsin D proteinase digestion site, and rRTA. In this work, the authors compared the cytotoxicity of the rIT with the chemical linked IT, evidencing that only the latter was toxic in target cells [115]. About 20 ricin-ITs have been tested in Phase I, II, and III clinical trials to treat patients with tumors, either hematological or solid, transplant rejection, and GvHD. The first Phase I clinical trial, giving promising results, was conducted by Spitler et al. in 1987 (Figure 1), by treating 22 metastatic malignant melanoma patients with an IT composed by murine monoclonal anti-melanoma antibody coupled to RTA (XOMAZYME-MEL) [116–118]. Additionally, ITs were also exploited for the treatment of autoimmune diseases. Indeed, anti-CD5/RTA was the first IT used in clinical trials for therapy of autoimmune diseases, such as rheumatoid arthritis, systemic lupus erythematosus, and insulin-dependent diabetes mellitus [119,120]. The advantages and limitations of ricin containing ITs for cancer therapy were recently discussed together with strategies for reducing the immunogenicity of recombinant ITs [121,122].

A different new approach consists of nanoparticle construction, in which ricin is genetically fused to carrier peptides that are able not only to recognize specific cellular target, but also to auto assemble, as stable nanoparticles, thus increasing the toxin-concentration into the targeted site [109,123,124].

6. Conclusions

In conclusion, ricin is a highly cytotoxic plant protein and has been of great utility to develop a number of anti-cancer immunotoxins. Ricin, and of some other RIPs, are able to act on multiple molecular targets inside the cell, thus triggering different death pathways; this makes such proteins more attractive for cancer treatment than conventional chemotherapy, in which one of the major problems is the rise of resistant cells [67,125]. However, ricin-containing ITs have also been shown to exhibit many limitations, such as unspecific toxicity, organ toxicity (mainly liver, kidney, and vasculature), immunogenicity, fast removal from blood stream, and lysosomal degradation inside cells. As a result, despite of the significant efforts made over the past few years, ricin as therapeutic agent has not achieved much impact at the clinical level. The challenge is still open, and frontline research is directed towards recombinant immunotoxins and nanocarriers, or towards other novel techniques such as the vector-driven expression of active plant toxin genes in tumor cells [109,126].

Although ricin is not toxic enough to hypothesize a use over a wide area for terrorist purposes, the availability of castor beans and the quite simple procedure for rough ricin purification have stimulated the interest of criminals and terrorists for small-scale biocrimes. This justifies researchers' efforts to obtain faster and more sensitive ricin detection tests. Furthermore, the study of inhibiting or neutralizing molecules and the timing of clinical events following ricin intoxication could lead to the definition of one or more validated therapies.

Finally, the use of castor bean derivatives should be carefully monitored because of the potential presence of active ricin. In fact, the large use of these products in agriculture, without an effective ricin inactivation, has already caused several cases of animal intoxication, and can be hazardous for human health.

Funding: This work was supported by funds for selected research topics of Alma Mater Studiorum—University of Bologna and by the Pallotti Legacies for Cancer Research.

Conflicts of Interest: The authors declare no conflict of interest.

References

1. D'Errico, F.; Backwell, L.; Villa, P.; Degano, I.; Lucejko, J.J.; Bamford, M.K.; Higham, T.F.; Colombini, M.P.; Beaumont, P.B. Early evidence of San material culture represented by organic artifacts from Border Cave, South Africa. *Proc. Natl. Acad. Sci. USA* **2012**, *109*, 13214–13219. [CrossRef] [PubMed]
2. Ebers, G. *Papyros Ebers: Das Hermetische Buch über die Arzneimittel der Alten Äegypter*; Hinrichs, J.C., Ed.; Wilhelm Engelmann: Leipzig, Germany, 1875.
3. Leake, C.D. *The Old Egyptian Medical Papyri*; University of Kansas Press: Lawrence, KS, USA, 1952.
4. Carpenter, S.; Rigaud, M.; Barile, M.; Priest, T.J.; Perez, L.; Ferguson, J.B. *An Interlinear Transliteration and English Translation of Portions of the Ebers Papyrus Possibly Having to do with Diabetes Mellitus*; Bard College: Annandale-on-Hudson, NY, USA, 1998.
5. Totelin, L.M.V. *Hippocratic Recipes: Oral and Written Transmission of Pharmacological Knowledge in Fifth- and Fourth-Century Greece*; Brill: Leiden, The Netherlands; Boston, MA, USA, 2009.
6. Gunther, R.T. *The Greek Herbal of Dioscorides/Illustrated by a Byzantine, A.D. 512; Englished by John Goodyer, A.D. 1655; Edited and First Printed, A.D. 1933*; Gunther, R.W.T., Ed.; Oxford University Press: New York, NY, USA, 1934.
7. Bostock, J.; Riley, H.T. *The Natural History of Pliny*, 1st ed.; Taylor and Francis: London, UK, 1855.
8. Scarpa, A.; Guerci, A. Various uses of the castor oil plant (*Ricinus communis* L.). A review. *J. Ethnopharmacol.* **1982**, *5*, 117–137. [CrossRef]
9. Ladda, P.L.; Kamthane, R.B. *Ricinus communis* (castor): An overview. *Int. J. Res. Pharmacol. Pharmacother.* **2014**, *3*, 136–144.
10. Quattrocchi, U. *CRC World Dictionary of Medicinal and Poisonous Plants: Common Names, Scientific Names, Eponyms, Synonyms, and Etymology*; Routledge: Abingdon-on-Thames, UK, 2012.
11. Polito, L.; Bortolotti, M.; Maiello, S.; Battelli, M.G.; Bolognesi, A. Plants producing Ribosome-Inactivating Proteins in traditional medicine. *Molecules* **2016**, *21*, 1560. [CrossRef] [PubMed]

12. Tunaru, S.; Althoff, T.F.; Nüsing, R.M.; Diener, M.; Offermanns, S. Castor oil induces laxation and uterus contraction via ricinoleic acid activating prostaglandin EP3 receptors. *Proc. Natl. Acad. Sci. USA* **2012**, *109*, 9179–9184. [CrossRef] [PubMed]

13. Adeniyi, O.S.; Omale, J.; Omeje, S.C.; Edino, V.O. Antidiarrheal activity of hexane extract of Citrus limon peel in an experimental animal model. *J. Integr. Med.* **2017**, *15*, 158–164. [CrossRef]

14. Battelli, M.G. Cytotoxicity and toxicity to animals and humans of ribosome-inactivating proteins. *Mini Rev. Med. Chem.* **2004**, *4*, 513–521. [CrossRef]

15. Barbieri, L.; Battelli, M.G.; Stirpe, F. Ribosome-inactivating proteins from plants. *Biochim. Biophys. Acta* **1993**, *1154*, 237–282. [CrossRef]

16. Dixson, T. *Ricinus communis. Aust. Med. Gazz.* **1887**, *6*, 137–138.

17. Hartmut, F. The ricin story. *Adv. Lectin Res.* **1988**, *1*, 10–25.

18. Stillmark, H. Über Ricin, ein Giftiges Ferment aus den Samen von *Ricinus communis* L. und Einigen Anderen Euphorbiaceen. Ph.D. Thesis, University of Dorpat, Dorpat, Estonia, 1888.

19. Takahashi, T.; Funatsu, G.; Funatsu, M. Biochemical studies on castor bean hemagglutinin. I. Separation and purification. *J. Biochem.* **1962**, *51*, 288–292. [CrossRef] [PubMed]

20. Ehrlich, P. Experimentelle Untersuchungen über Immunität I. Über Ricin. *DMW-Deutsche Med. Wochenschr.* **1891**, *17*, 976–979. [CrossRef]

21. Ehrlich, P. Experimentelle Untersuchungen über Immunität II. Über Abrin. *DMW-Deutsche Med. Wochenschr.* **1891**, *17*, 1218–1219. [CrossRef]

22. Silverstein, A.M. Paul Ehrlich: The founding of pediatric immunology. *Cell Immunol.* **1996**, *174*, 1–6. [CrossRef] [PubMed]

23. Ehrlich, P. Partial Cell Functions. In *Nobel Lectures, Physiology or Medicine 1901–1921*; Elsevier Publishing Company: Amsterdam, The Netherlands, 1967.

24. Lin, J.Y.; Tserng, K.Y.; Chen, C.C.; Lin, L.T.; Tung, T.C. Abrin & Ricin: New anti-tumour substances. *Nature* **1970**, *227*, 292–293. [PubMed]

25. Lin, J.Y.; Liu, K.; Chen, C.C.; Tung, T.C. Effect of crystalline ricin on the biosynthesis of protein, RNA, and DNA in experimental tumor cells. *Cancer Res.* **1971**, *31*, 921–924. [PubMed]

26. Olsnes, S.; Pihl, A. Different biological properties of the two constituent peptide chains of ricin, a toxic protein inhibiting protein synthesis. *Biochemistry* **1973**, *12*, 3121–3126. [CrossRef]

27. Olsnes, S.; Pihl, A. Inhibition of peptide chain elongation. *Nature* **1972**, *238*, 459–461. [CrossRef]

28. Funatsu, G.; Yoshitake, S.; Funatsu, M. Primary Structure of Ile Chain of Ricin D. *Agric. Biol. Chem.* **1978**, *42*, 501–503. [CrossRef]

29. Funatsu, G.; Kimura, M.; Funatsu, M. Primary Structure of Ala Chain of Ricin D. *Agric. Biol. Chem.* **1979**, *43*, 2221–2224. [CrossRef]

30. Montfort, W.; Villafranca, J.E.; Monzingo, A.F.; Ernst, S.R.; Katzin, B.; Rutenber, E.; Xuong, N.H.; Hamlin, R.; Robertus, J.D. The three-dimensional structure of ricin at 2.8 A. *J. Biol. Chem.* **1987**, *262*, 5398–5403. [PubMed]

31. Rutenber, E.; Katzin, B.J.; Ernst, S.; Collins, E.J.; Mlsna, D.; Ready, M.P.; Robertus, J.D. Crystallographic refinement of ricin to 2.5 A. *Proteins* **1991**, *10*, 240–250. [CrossRef] [PubMed]

32. Katzin, B.J.; Collins, E.J.; Robertus, J.D. Structure of ricin A-chain at 2.5 A. *Proteins* **1991**, *10*, 251–259. [CrossRef] [PubMed]

33. Rutenber, E.; Robertus, J.D. Structure of ricin B-chain at 2.5 A resolution. *Proteins* **1991**, *10*, 260–269. [CrossRef] [PubMed]

34. Weston, S.A.; Tucker, A.D.; Thatcher, D.R.; Derbyshire, D.J.; Pauptit, R.A. X-ray structure of recombinant ricin A-chain at 1.8 A resolution. *J. Mol. Biol.* **1994**, *244*, 410–422. [CrossRef] [PubMed]

35. Vater, C.A.; Bartle, L.M.; Leszyk, J.D.; Lambert, J.M.; Goldmacher, V.S. Ricin A chain can be chemically cross-linked to the mammalian ribosomal proteins L9 and L10e. *J. Biol. Chem.* **1995**, *270*, 12933–12940. [CrossRef] [PubMed]

36. Monzingo, A.F.; Robertus, J.D. X-ray analysis of substrate analogs in the ricin A-chain active site. *J. Mol. Biol.* **1992**, *227*, 1136–1145. [CrossRef]

37. Ready, M.P.; Kim, Y.; Robertus, J.D. Site-directed mutagenesis of ricin A-chain and implications for the mechanism of action. *Proteins* **1991**, *10*, 270–278. [CrossRef]

38. Olsnes, S.; Pihl, A. Ricin-a potent inhibitor of protein synthesis. *FEBS Lett.* **1972**, *20*, 327–329. [CrossRef]

39. Sperti, S.; Montanaro, L.; Mattioli, A.; Stirpe, F. Inhibition by ricin of protein synthesis in vitro: 60 S ribosomal subunit as the target of the toxin. *Biochem. J.* **1973**, *136*, 813–815. [CrossRef]

40. Montanaro, L.; Sperti, S.; Stirpe, F. Inhibition by ricin of protein synthesis in vitro. Ribosomes as the target of the toxin. *Biochem. J.* **1973**, *136*, 677–683. [CrossRef] [PubMed]

41. Sperti, S.; Montanaro, L.; Mattioli, A.; Testoni, G. Relationship between elongation factor I- and elongation factor II—Dependent guanosine triphosphatase activities of ribosomes. Inhibition of both activities by ricin. *Biochem. J.* **1975**, *148*, 447–451. [CrossRef] [PubMed]

42. Benson, S.; Olsnes, S.; Pihl, A.; Skorve, J.; Abraham, A.K. On the mechanism of protein-synthesis inhibition by abrin and ricin. Inhibition of the GTP-hydrolysis site on the 60-S ribosomal subunit. *Eur. J. Biochem.* **1975**, *59*, 573–580. [CrossRef] [PubMed]

43. Olsnes, S.; Fernandez-Puentes, C.; Carrasco, L.; Vazquez, D. Ribosome inactivation by the toxic lectins abrin and ricin. Kinetics of the enzymic activity of the toxin A-chains. *Eur. J. Biochem.* **1975**, *60*, 281–288. [CrossRef] [PubMed]

44. Barbieri, L.; Stirpe, F. Ribosome-inactivating proteins from plants: Properties and possible uses. *Cancer Surv.* **1982**, *1*, 489–520.

45. Bolognesi, A.; Bortolotti, M.; Maiello, S.; Battelli, M.G.; Polito, L. Ribosome-Inactivating Proteins from Plants: A Historical Overview. *Molecules* **2016**, *21*, 1627. [CrossRef]

46. Endo, Y.; Tsurugi, K. RNA N-glycosidase activity of ricin A-chain. Mechanism of action of the toxic lectin ricin on eukaryotic ribosomes. *J. Biol. Chem.* **1987**, *262*, 8128–8130.

47. Endo, Y.; Mitsui, K.; Motizuki, M.; Tsurugi, K. The mechanism of action of ricin and related toxic lectins on eukaryotic ribosomes. The site and the characteristics of the modification in 28 S ribosomal RNA caused by the toxins. *J. Biol. Chem.* **1987**, *262*, 5908–5912.

48. Barbieri, L.; Valbonesi, P.; Bonora, E.; Gorini, P.; Bolognesi, A.; Stirpe, F. Polynucleotide:adenosine glycosidase activity of ribosome-inactivating proteins: Effect on DNA, RNA and poly(A). *Nucleic Acids Res.* **1997**, *25*, 518–522. [CrossRef]

49. Barbieri, L.; Bolognesi, A.; Valbonesi, P.; Polito, L.; Olivieri, F.; Stirpe, F. Polynucleotide: Adenosine glycosidase activity of immunotoxins containing ribosome-inactivating proteins. *J. Drug Target.* **2000**, *8*, 281–288. [CrossRef]

50. Barbieri, L.; Brigotti, M.; Perocco, P.; Carnicelli, D.; Ciani, M.; Mercatali, L.; Stirpe, F. Ribosome-inactivating proteins depurinate poly(ADP-ribosyl)ated poly(ADP-ribose) polymerase and have transforming activity for 3T3 fibroblasts. *FEBS Lett.* **2003**, *538*, 178–182. [CrossRef]

51. Brigotti, M.; Alfieri, R.; Sestili, P.; Bonelli, M.; Petronini, P.G.; Guidarelli, A.; Barbieri, L.; Stirpe, F.; Sperti, S. Damage to nuclear DNA induced by Shiga toxin 1 and ricin in human endothelial cells. *FASEB J.* **2002**, *16*, 365–372. [CrossRef] [PubMed]

52. Sandvig, K.; Olsnes, S.; Pihl, A. Kinetics of binding of the toxic lectins abrin and ricin to surface receptors of human cells. *J. Biol. Chem.* **1976**, *251*, 3977–3984. [PubMed]

53. van Deurs, B.; Petersen, O.W.; Sundan, A.; Olsnes, S.; Sandvig, K. Receptor-mediated endocytosis of a ricin-colloidal gold conjugate in Vero cells: Intracellular routing to vacuolar and tubulo-vesicular portions of the endosomal system. *Exp. Cell Res.* **1985**, *159*, 287–304. [CrossRef]

54. Moya, M.; Dautry-Varsat, A.; Goud, B.; Louvard, D.; Boquet, P. Inhibition of coated pit formation in Hep2 cells blocks the cytotoxicity of diphtheria toxin but not that of ricin. *J. Cell Biol.* **1985**, *101*, 548–559. [CrossRef] [PubMed]

55. van Deurs, B.; Sandvig, K.; Petersen, O.W.; Olsnes, S.; Simons, K.; Griffiths, G. Routing of internalised ricin and ricin conjugates to the Golgi complex. *J. Cell Biol.* **1986**, *102*, 37–47. [CrossRef] [PubMed]

56. Sandvig, K.; van Deurs, B. Endocytosis, intracellular transport, and cytotoxic action of Shiga toxin and ricin. *Physiol. Rev.* **1996**, *76*, 949–966. [CrossRef]

57. Cosson, P.; Letourneur, F. Coatamer interaction with di-lysine endoplasmic reticulum retention motifs. *Science* **1994**, *263*, 1629–1631. [CrossRef]

58. Chen, A.; Hu, T.; Mikoryak, C.; Draper, R.K. Retrograde transport of protein toxins under conditions of COPI dysfunction. *Biochim. Biophys. Acta* **2002**, *1589*, 124–139. [CrossRef]

59. Wesche, J.; Rapak, A.; Olsnes, S. Dependence of ricin toxicity on translocation of the toxin A-chain from the endoplasmic reticulum to the cytosol. *J. Biol. Chem.* **1999**, *274*, 34443–34449. [CrossRef]

60. Sandvig, K.; Bergan, J.; Kavaliauskiene, S.; Skotland, T. Lipid requirements for entry of protein toxins into cells. *Prog. Lipid Res.* **2014**, *54*, 1–13. [CrossRef] [PubMed]

61. Roberts, L.M.; Smith, D.C. Ricin: The endoplasmic reticulum connection. *Toxicon* **2004**, *44*, 469–472. [CrossRef] [PubMed]

62. Eiklid, K.; Olsnes, S.; Pihl, A. Entry of lethal doses of abrin, ricin and modeccin into the cytosol of HeLa cells. *Exp. Cell Res.* **1980**, *126*, 321–326. [CrossRef]

63. Harper, C.G.; Gonatas, J.O.; Mizutani, T.; Gonatas, N.K. Retrograde transport and effects of toxic ricin in the autonomic nervous system. *Lab. Investig.* **1980**, *42*, 396–404. [PubMed]

64. Polito, L.; Bortolotti, M.; Farini, V.; Battelli, M.G.; Barbieri, L.; Bolognesi, A. Saporin induces multiple death pathways in lymphoma cells with different intensity and timing as compared to ricin. *Int. J. Biochem. Cell Biol.* **2009**, *41*, 1055–1061. [CrossRef]

65. Tazzari, P.L.; Bolognesi, A.; de Totero, D.; Falini, B.; Lemoli, R.M.; Soria, M.R.; Pileri, S.; Gobbi, M.; Stein, H.; Flenghi, L.; et al. Ber-H2 (anti-CD30)-saporin immunotoxin: A new tool for the treatment of Hodgkin's disease and CD30+ lymphoma: In vitro evaluation. *Br. J. Haematol.* **1992**, *81*, 203–211. [CrossRef]

66. Ngo, T.N.; Nguyen, T.T.; Bui, D.T.-T.; Hoang, N.T.-M.; Nguyen, T.D. Effects of ricin extracted from seeds of the castor bean (*Ricinus communis*) on cytotoxicity and tumorigenesis of melanoma cells. *Biomed. Res. Ther.* **2016**, *3*, 633–644.

67. Polito, L.; Bortolotti, M.; Pedrazzi, M.; Mercatelli, D.; Battelli, M.G.; Bolognesi, A. Apoptosis and necroptosis induced by stenodactylin in neuroblastoma cells can be completely prevented through caspase inhibition plus catalase or necrostatin-1. *Phytomedicine* **2016**, *23*, 32–41. [CrossRef]

68. Roxas-Duncan, V.I.; Smith, L.A. Of Beans and Beads: Ricin and abrin in bioterrorism and biocrime. *J. Bioterr. Biodef.* **2012**, *S2*, 002. [CrossRef]

69. Whitfield, S.J.C.; Griffiths, G.D.; Jenner, D.C.; Gwyther, R.J.; Stahl, F.M.; Cork, L.J.; Holley, J.L.; Green, A.C.; Clark, G.C. Production, characterisation and testing of an ovine antitoxin against ricin; efficacy, potency and mechanisms of action. *Toxins* **2017**, *9*, 329. [CrossRef]

70. Audi, J.; Belson, M.; Patel, M.; Schier, J.; Osterloh, J. Ricin poisoning-a comprehensive review. *JAMA* **2005**, *294*, 2342–2351. [CrossRef] [PubMed]

71. Cook, D.L.; David, J.; Griffiths, G.D. Retrospective identification of ricin in animal tissues following administration by pulmonary and oral routes. *Toxicology* **2006**, *223*, 61–70. [CrossRef] [PubMed]

72. Roy, C.J.; Song, K.; Sivasubramani, S.K.; Gardner, D.J.; Pincus, S.H. Animal models of ricin toxicosis. *Curr. Top. Microbiol. Immunol.* **2012**, *357*, 243–257. [PubMed]

73. He, X.; Carter, J.M.; Brandon, D.L.; Cheng, L.W.; McKeon, T.A. Application of a real time polymerase chain reaction method to detect castor toxin contamination in fluid milk and eggs. *J. Agric. Food Chem.* **2007**, *55*, 6897–6902. [CrossRef] [PubMed]

74. Olsnes, S. The history of ricin, abrin and related toxins. *Toxicon* **2004**, *44*, 361–370. [CrossRef]

75. Wedin, G.P.; Neal, J.S.; Everson, G.W.; Krenzelok, E.P. Castor bean poisoning. *Am. J. Emerg. Med.* **1986**, *4*, 259–261. [CrossRef]

76. Worbs, S.; Köhler, K.; Pauly, D.; Avondet, M.A.; Schaer, M.; Dorner, M.B.; Dorner, B.G. *Ricinus communis* intoxications in human and veterinary medicine-a summary of real cases. *Toxins* **2011**, *3*, 1332–1372. [CrossRef]

77. Grimshaw, B.; Wennike, N.; Dayer, M. Ricin poisoning: A case of internet-assisted parasuicide. *Br. J. Hosp. Med.* **2013**, *74*, 532–533. [CrossRef]

78. Bradberry, S. Ricin and abrin. *Medicine* **2016**, *44*, 109–110. [CrossRef]

79. Bradberry, S.M.; Dickers, K.J.; Rice, P.; Griffiths, G.D.; Vale, J.A. Ricin poisoning. *Toxicol. Rev.* **2003**, *22*, 65–70. [CrossRef]

80. OPCW-Organisation for the Prohibition of Chemical Weapons. Ricin Fact Sheet 2014. Available online: https://www.opcw.org/sites/default/files/documents/SAB/en/sab-21-wp05_e_.pdf (accessed on 15 April 2019).

81. Pincus, S.H.; Bhaskaran, M.; Brey, R.N., 3rd; Didier, P.J.; Doyle-Meyers, L.A.; Roy, C.J. Clinical and Pathological Findings Associated with Aerosol Exposure of Macaques to Ricin Toxin. *Toxins* **2015**, *7*, 2121–2133. [CrossRef]

82. Franz, D.R.; Jaax, N.K. Ricin Toxin. In *Medical Aspects of Chemical and Biological Warfare*; Sidell, F.R., Takafuji, E.T., Franz, D.R., Eds.; Walter Reed Army Medical Center, Borden Institute: Washington, DC, USA, 1997; Volume 3, pp. 631–642.

83. Millard, C.; LeClaire, R. Ricin and related toxins: Review and perspective. In *Chemical Warfare Agents: Chemistry, Pharmacology, Toxicology, and Therapeutics*, 2nd ed.; Brian, J., Lukey, J.A., Romano, J.A., Jr., Harry Salem, R., Lukey, B.J., Salem, H., Eds.; CRC Press, Taylor & Francis Group: Boca Raton, FL, USA, 2007; pp. 424–452.

84. Crompton, R.; Gall, D. Georgi Markov-death in a pellet. *Med. Leg. J.* **1980**, *48*, 51–62. [CrossRef] [PubMed]

85. Kortepeter, M.G.; Parker, G.W. Potential biological weapons threats. *Emerg. Infect. Dis.* **1999**, *5*, 523–527. [CrossRef] [PubMed]

86. Mubofu, E.B. Castor oil as a potential renewable resource for the production of functional materials. *Sustain. Chem. Process.* **2016**, *4*, 11. [CrossRef]

87. Ordman, D. An outbreak of bronchial asthma in South Africa, affecting more than 200 persons, caused by castor bean dust from an oil processing factory. *Int. Arch. Allergy Appl. Immunol.* **1955**, *7*, 10–24. [CrossRef]

88. Panzani, R.C. Respiratory castor bean dust allergy in the south of France with special reference to Marseilles. *Int. Arch. Allergy Appl. Immunol.* **1957**, *11*, 224–236. [CrossRef] [PubMed]

89. Weiss, E.A. Castor. In *Oilseed Crops*, 2nd ed.; Blackwell Science Ltd.: Oxford, UK, 2000; pp. 31–36.

90. Vashst, D.; Amhad, M. Statistical analysis of diesel engine performance for castor and jatropha biodiesel-blended fuel. *IJAME* **2014**, *10*, 2155–2169. [CrossRef]

91. Falk, S. World Outlook for Castor Oil 2018. In Proceedings of the Global Castor Conference 2018, Ahmedabad, India, 23–24 February 2018.

92. Food and Agriculture Organization of United States (FAO). 2017. Available online: http://www.fao.org/faostat/en/#data/QC (accessed on 15 April 2019).

93. Patel, V.R.; Dumancas, G.G.; Kasi Viswanath, L.C.; Maples, R.; Subong, B.J. Castor oil: Properties, uses, and optimization of processing parameters in commercial production. *Lipid Insights* **2016**, *9*, 1–12. [CrossRef]

94. Alexander, J.; Benford, D.; Cockburn, A.; Cravedi, J.-P.; Dogliotti, E.; Di Domenico, A.; Férnandez-Cruz, M.L.; Fürst, P.; Fink-Gremmels, J.; Galli, C.L.; et al. Scientific Opinion of the Panel on Contaminants in the Food Chain on a request from the European Commission on ricin (from *Ricinus communis*) as undesirable substances in animal feed. *EFSA J.* **2008**, *726*, 9–38. [CrossRef]

95. Anandan, S.; Kumar, G.; Ghosh, J.; Ramachandra, K. Effect of different physical and chemical treatments on detoxification of ricin in castor cake. *Anim. Feed Sci. Tech.* **2004**, *120*, 159–168. [CrossRef]

96. Smallshaw, J.E.; Vitetta, E.S. Ricin vaccine development. *Curr. Top. Microbiol. Immunol.* **2012**, *357*, 259–272. [PubMed]

97. Rainey, G.J.; Young, J.A. Antitoxins: Novel strategies to target agents of bioterrorism. *Nat. Rev. Microbiol.* **2004**, *2*, 721–726. [CrossRef] [PubMed]

98. Stechmann, B.; Bai, S.K.; Gobbo, E.; Lopez, R.; Merer, G.; Pinchard, S.; Panigai, L.; Tenza, D.; Raposo, G.; Beaumelle, B.; et al. Inhibition of retrograde transport protects mice from lethal ricin challenge. *Cell* **2010**, *141*, 231–242. [CrossRef] [PubMed]

99. Barbier, J.; Bouclier, C.; Johannes, L.; Gillet, D. Inhibitors of the cellular trafficking of ricin. *Toxins* **2012**, *4*, 15–27. [CrossRef] [PubMed]

100. Gal, Y.; Mazor, O.; Alcalay, R.; Seliger, N.; Aftalion, M.; Sapoznikov, A.; Falach, R.; Kronman, C.; Sabo, T. Antibody/doxycycline combined therapy for pulmonary ricinosis: Attenuation of inflammation improves survival of ricin-intoxicated mice. *Toxicol. Rep.* **2014**, *1*, 496–504. [CrossRef] [PubMed]

101. O'Hara, J.M.; Whaley, K.; Pauly, M.; Zeitlin, L.C.; Mantis, N. Plant-based expression of a partially humanized neutralizing monoclonal IgG directed against an immunodominant epitope on the ricin toxin A subunit. *Vaccine* **2012**, *30*, 1239–1243. [CrossRef] [PubMed]

102. Sully, E.K.; Whaley, K.J.; Bohorova, N.; Goodman, C.; Kim, D.H.; Pauly, M.H.; Velasco, J.; Hiatt, E.; Morton, J.; Swope, K.; et al. Chimeric plantibody passively protects mice against aerosolized ricin challenge. *Clin. Vaccine Immunol.* **2014**, *21*, 777–782. [CrossRef]

103. Van Slyke, G.; Sully, E.K.; Bohorova, N.; Bohorov, O.; Kim, D.; Pauly, M.; Whaley, K.; Zeitlin, L.; Mantis, N.J. Humanized monoclonal antibody that passively protects mice 3 against systemic and intranasal ricin toxin challenge. *Clin. Vaccine Immunol.* **2016**, *23*, 795–799. [CrossRef]

104. Pratt, T.S.; Pincus, S.H.; Hale, M.L.; Moreira, A.L.; Roy, C.J.; Tchou-Wong, K.-M. Oropharyngeal aspiration of ricin as a lung challenge model for evaluation of the therapeutic index of antibodies against ricin A chain for post-exposure treatment. *Exp. Lung Res.* **2007**, *33*, 459–481. [CrossRef]

105. Noy-Porat, T.; Alcalay, R.; Epstein, E.; Sabo, T.; Kronman, C. Extended therapeutic window for post exposure treatment of ricin intoxication conferred by the use of high-affinity antibodies. *Toxicon* **2017**, *127*, 100–105. [CrossRef]
106. Noy-Porat, T.; Rosenfeld, R.; Ariel, N.; Epstein, E.; Alcalay, R.; Zvi, A.; Kronman, C.; Ordentlich, A.; Mazor, O. Isolation of anti-ricin protective antibodies exhibiting high affinity from immunized non-human primates. *Toxins* **2016**, *8*, 64. [CrossRef] [PubMed]
107. Moolten, F.; Zajdel, S.; Cooperband, S. Immunotherapy of experimental animal tumors with antitumor antibodies conjugated to diphtheria toxin or ricin. *Ann. N. Y. Acad. Sci.* **1976**, *277*, 690–699. [CrossRef] [PubMed]
108. Gilabert-Oriol, R.; Weng, A.; Mallinckrodtm, B.; Melzig, M.F.; Fuchs, H.; Thakur, M. Immunotoxins constructed with ribosome-inactivating proteins and their enhancers: A lethal cocktail with tumor specific efficacy. *Curr. Pharm. Des.* **2014**, *20*, 6584–6643. [CrossRef] [PubMed]
109. Tyagi, N.; Tyagi, M.; Pachauri, M.; Ghosh, P.C. Potential therapeutic applications of plant toxin-ricin in cancer: Challenges and advances. *Tumour Biol.* **2015**, *36*, 8239–8246. [CrossRef] [PubMed]
110. Blythman, H.E.; Casellas, P.; Gros, O.; Gros, P.; Jansen, F.K.; Paolucci, F.; Pau, B.; Vidal, H. Immunotoxins: Hybrid molecules of monoclonal antibodies and a toxin subunit specifically kill tumour cells. *Nature* **1981**, *290*, 145–146. [CrossRef]
111. Weil-Hillman, G.; Runge, W.; Jansen, F.K.; Vallera, D.A. Cytotoxic effect of anti-Mr 67,000 protein immunotoxins on human tumors in a nude mouse model. *Cancer Res.* **1985**, *45*, 1328–1336.
112. Ghetie, V.; Ghetie, M.A.; Uhr, J.W.; Vitetta, E.S. Large scale preparation of immunotoxins constructed with the Fab' fragment of IgG1 murine monoclonal antibodies and chemically deglycosylated ricin A chain. *J. Immunol. Methods* **1988**, *112*, 267–277. [CrossRef]
113. Ghetie, M.A.; Richardson, J.; Tucker, T.; Jones, D.; Uhr, J.W.; Vitetta, E.S. Antitumor activity of Fab' and IgG-anti-CD22 immunotoxins in disseminated human B lymphoma grown in mice with severe combined immunodeficiency disease: Effect on tumor cells in extranodal sites. *Cancer Res.* **1991**, *51*, 5876–5880.
114. FitzGerald, D.J.; Bjorn, M.J.; Ferris, R.J.; Winkelhake, J.L.; Frankel, A.E.; Hamilton, T.C.; Ozols, R.F.; Willingham, M.C.; Pastan, I. Antitumor activity of an immunotoxin in a nude mouse model of human ovarian cancer. *Cancer Res.* **1987**, *47*, 1407–1410.
115. Wang, D.; Li, Q.; Hudson, W.; Berven, E.; Uckun, F.; Kersey, J.H. Generation and characterization of an anti-CD19 single-chain Fv immunotoxin composed of C-terminal disulfide-linked dgRTA. *Bioconj. Chem.* **1997**, *8*, 878–884. [CrossRef]
116. Hertler, A.A.; Splitler, L.; Frankel, A. Humoral immune response to a ricin A chain immunotoxin in patients with metastatic melanoma. *Cancer Drug Deliv.* **1987**, *4*, 245–253. [CrossRef] [PubMed]
117. Spitler, L.E.; del Rio, M.; Khentigan, A.; Wedel, N.I.; Brophy, N.A.; Miller, L.L.; Harkonen, W.S.; Rosendorf, L.L.; Lee, H.M.; Mischak, R.P.; et al. Therapy of patients with malignant melanoma using a monoclonal antimelanoma antibody-ricin A chain immunotoxin. *Cancer Res.* **1987**, *47*, 1717–1723. [PubMed]
118. Mischak, R.P.; Foxall, C.; Rosendorf, L.L.; Knebel, K.; Scannon, P.J.; Spitler, L.E. Human antibody responses to components of the monoclonal antimelanoma antibody ricin A chain immunotoxin Xoma-Zyme-MEL. *Mol. Biother.* **1990**, *2*, 104–109. [PubMed]
119. Strand, V.; Lipsky, P.E.; Cannon, G.W.; Calabrese, L.H.; Wiesenhutter, C.; Cohen, S.B.; Olsen, N.J.; Lee, M.L.; Lorenz, T.J.; Nelson, B. Effects of administration of an anti-CD5 plus immunoconjugate in rheumatoid arthritis. Results of two phase II studies. The CD5 plus rheumatoid arthritis investigators group. *Arthritis Rheum.* **1993**, *36*, 620–630. [CrossRef] [PubMed]
120. Stafford, F.J.; Fleisher, T.A.; Lee, G.; Brown, M.; Strand, V.; Austin, H.A.; Balow, J.E.; Klippel, J.H. A pilot study of anti-CD5 ricin A chain immunoconjugate in systemic lupus erythematosus. *J. Rheumatol.* **1994**, *21*, 2068–2070. [PubMed]
121. Polito, L.; Djemil, A.; Bortolotti, M. Plant Toxin-Based Immunotoxins for Cancer Therapy: A Short Overview. *Biomedicines* **2016**, *4*, E12. [CrossRef] [PubMed]
122. Akbari, B.; Farajnia, S.; Ahdi Khosroshahi, S.; Safari, F.; Yousefi, M.; Dariushnejad, H.; Rahbarnia, L. Immunotoxins in cancer therapy: Review and update. *Int. Rev. Immunol.* **2017**, *36*, 207–219. [CrossRef] [PubMed]

123. Ashley, C.E.; Carnes, E.C.; Phillips, G.K.; Durfee, P.N.; Buley, M.D.; Lino, C.A.; Padilla, D.P.; Phillips, B.; Carter, M.B.; Willman, C.L.; et al. Cell-specific delivery of diverse cargos by bacteriophage MS2 virus-like particles. *ACS Nano* **2011**, *5*, 5729–5745. [CrossRef] [PubMed]
124. Díaz, R.; Pallarès, V.; Cano-Garrido, O.; Serna, N.; Sánchez-García, L.; Falgàs, A.; Pesarrodona, M.; Unzueta, U.; Sánchez-Chardi, A.; Sánchez, J.M.; et al. Selective CXCR4(+) cancer cell targeting and potent antineoplastic effect by a nanostructured version of recombinant ricin. *Small* **2018**, *14*, 1800665. [CrossRef]
125. Polito, L.; Mercatelli, D.; Bortolotti, M.; Maiello, S.; Djemil, A.; Battelli, M.G.; Bolognesi, A. Two Saporin-Containing Immunotoxins Specific for CD20 and CD22 Show Different Behavior in Killing Lymphoma Cells. *Toxins* **2017**, *9*, 182. [CrossRef]
126. Zhang, Y.H.; Wang, Y.; Yusufali, A.H.; Ashby, F.; Zhang, D.; Yin, Z.F.; Aslanidi, G.V.; Srivastava, A.; Ling, C.Q.; Ling, C. Cytotoxic genes from traditional Chinese medicine inhibit tumor growth both in vitro and in vivo. *J. Integr. Med.* **2014**, *12*, 483–494. [CrossRef]

 *toxins*

Review

Intracellular Transport and Cytotoxicity of the Protein Toxin Ricin

Natalia Sowa-Rogozińska [1], Hanna Sominka [1], Jowita Nowakowska-Gołacka [1], Kirsten Sandvig [2,3] and Monika Słomińska-Wojewódzka [1,*]

[1] Department of Medical Biology and Genetics, Faculty of Biology, University of Gdańsk, Wita Stwosza 59, 80-308 Gdańsk, Poland; natalia.sowa@phdstud.ug.edu.pl (N.S.-R.); hanna.sominka@phdstud.ug.edu.pl (H.S.); jowita.nowakowska@phdstud.ug.edu.pl (J.N.-G.)

[2] Department of Molecular Cell Biology, Institute for Cancer Research, Oslo University Hospital, 0379 Oslo, Norway; kirsten.sandvig@ibv.uio.no

[3] Faculty of Mathematics and Natural Sciences, Department of Biosciences, University of Oslo, 0316 Oslo, Norway

* Correspondence: monika.slominska@biol.ug.edu.pl; Tel.: +48-585-236-035

Received: 22 May 2019; Accepted: 14 June 2019; Published: 18 June 2019

Abstract: Ricin can be isolated from the seeds of the castor bean plant (*Ricinus communis*). It belongs to the ribosome-inactivating protein (RIP) family of toxins classified as a bio-threat agent due to its high toxicity, stability and availability. Ricin is a typical A-B toxin consisting of a single enzymatic A subunit (RTA) and a binding B subunit (RTB) joined by a single disulfide bond. RTA possesses an RNA N-glycosidase activity; it cleaves ribosomal RNA leading to the inhibition of protein synthesis. However, the mechanism of ricin-mediated cell death is quite complex, as a growing number of studies demonstrate that the inhibition of protein synthesis is not always correlated with long term ricin toxicity. To exert its cytotoxic effect, ricin A-chain has to be transported to the cytosol of the host cell. This translocation is preceded by endocytic uptake of the toxin and retrograde traffic through the trans-Golgi network (TGN) and the endoplasmic reticulum (ER). In this article, we describe intracellular trafficking of ricin with particular emphasis on host cell factors that facilitate this transport and contribute to ricin cytotoxicity in mammalian and yeast cells. The current understanding of the mechanisms of ricin-mediated cell death is discussed as well. We also comment on recent reports presenting medical applications for ricin and progress associated with the development of vaccines against this toxin.

Keywords: ricin; protein synthesis inhibition; apoptosis

Key Contribution: This review summarizes the current knowledge of ricin intracellular transport and mechanisms of its cytotoxicity.

---

1. Introduction

The toxin ricin is a naturally occurring, extremely toxic protein isolated from the seeds of the castor plant, *Ricinus communis*. This toxin was first described in 1888 by Peter Hermann Stillmark, who identified the active ingredient isolated from the castor seeds as a protein [1]. Interestingly, he described in his doctoral thesis, ricin's agglutinating properties, being the first person who defined and characterized a specific carbohydrate-binding protein, that later was called lectin [2]. At that time, the ability of extracts from *Ricinus communis* to agglutinate erythrocytes and to precipitate serum proteins was considered to be the mechanism behind the cytotoxic action of ricin. Later experiments showed that the toxicity and agglutination effects are separate properties of this toxin [3]. The structure of ricin was described by Olsnes and Pihl [4]. They also found that ricin exerts its cytotoxic effect by enzymatic action on eukaryotic ribosomes resulting in inhibition of protein synthesis [3–5]. Considering the mechanism of ricin cytotoxic activity, it was the very first identified RIP (ribosome-inactivating protein) assigned to the class II of this group of protein toxins [6–8].

Ricin belongs to the A-B family of protein toxins. The A-chain (RTA, ~32 kDa) is linked by a single disulfide bond to the lectin B-chain (RTB, ~34 kDa). Together they form ricin holotoxin (Figure 1). RTA possesses an enzymatic activity [6,7], whereas RTB binds to eukaryotic cell-surface glycoproteins and glycolipids [9]. Ricin A-chain specifically depurinates the α-sarcin-ricin loop (SRL) by hydrolyzing the N-glycosidic bond at adenine 4324 located at a GAGA hairpin of SRL of the 28S rRNA in the eukaryotic large ribosomal subunit [5], (for review see Refs. [8,10,11]). Interestingly, ricin does not remove an adenine from rRNA in whole *E. coli* ribosomes, thus genes coding for ricin could be expressed in *E. coli* [12]. It is considered that the sarcin-ricin loop is one the largest universally-conserved regions of the ribosome [13,14]. This highlights its importance in ribosome function. Indeed, SRL significantly influences the proper assembly of the functional structure of the 50S prokaryotic subunit [15], and it is highly probable that this loop fulfills a similar role in the large ribosomal subunit in eukaryotic cells. However, what is most important for ricin toxicity is that depurination of SRL prevents the binding of two crucial factors operating in the machinery of protein synthesis: the eukaryotic elongation factor 1 (eEF-1) and the elongation factor 2 (eEF-2) [9,16,17]. This blocks protein synthesis and is a prerequisite for the cytotoxic effect of ricin. A single ricin A-chain molecule is able to inactivate approximately 1500 ribosomes per minute [18,19]. It happens much faster than the cell can produce new ones [20]. Ricin's lethal dose in humans was estimated to be about 1.78 mg for an average adult [21]. However, its toxicity depends on the route of exposure. Inhalation is more potent than oral administration. The inhalation median lethal dose (LD50) is 3–5 µg/kg, while the oral LD50 is 20 mg/kg [22]. Due to ricin's high toxicity and stability, ease of production and good availability, it has been classified by the US Centers for Disease Control and Prevention (CDC) as a Category B Select Agent. Implementation of the Chemical Weapons Convention (CWC) in the national legislation of the 192 signatory countries (June 2017) makes undeclared ricin purification a global crime [23]. Despite the fact that ricin-mediated depurination of rRNA has been quite well described, other mechanisms involved in its cytotoxicity are not completely clarified. In fact, the inhibition of protein synthesis by ricin A-chain is not exclusively responsible for the cytotoxic effect of this toxin [24]. It has been demonstrated that ricin can induce apoptosis, cell membrane damage, membrane structure and function alteration, and release of cytokine inflammatory mediators [25–30]. In general, the inhibition of protein synthesis seems to precede apoptosis and be necessary for this event. It was, however, suggested that two different motifs present in ricin A-chain may be involved in ricin-mediated inhibition of protein synthesis and apoptosis [31,32] and that B-chain in human myeloid leukemia cells (U937) is able to induce apoptosis through its lectin activity without the contribution of the A-chain [33].

Elucidation of the entire mechanisms of ricin toxicity is crucial to fully utilize, but also to control all properties of this toxin. Ricin is being considered as one of the most toxic substances that exists. It can be used as a potential tool in bioterrorist attacks [34,35]. Thus, the development of effectively working antitoxin agents is of particular interest [36–39]. On the other hand, ricin conjugated with specific antibodies, other proteins, peptides or nanoparticles can be selectively directed to target cells. This ensures the possibility of a huge application of this toxin in medicine [40–44]. In this review, we describe the most important steps of ricin intracellular transport as well as diverse and complicated mechanisms of its action on cells. We also summarize the newest reports concerning the development of vaccines against ricin and biomedical applications of this toxin.

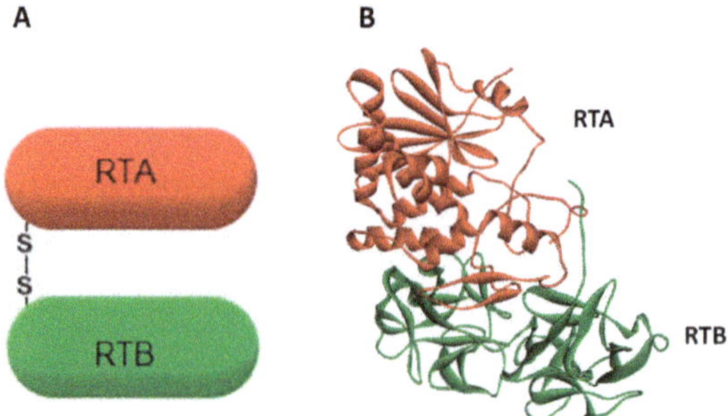

Figure 1. Schematic representation (**A**) and crystal structure (**B**) of the toxin ricin. The enzymatically-active subunit (A-chain) is marked in red, whereas the binding domain (B-chain) is presented in green. Both subunits are linked by a single disulfide bond. Crystal structure has been obtained from the PDB protein data bank (code 2AA1).

2. Intracellular Transport of Ricin

2.1. Uptake of Ricin into the Cell

Ricin B-chain recognizes complex types of carbohydrate receptors present on the surface of eukaryotic cells. These receptors contain either terminal N-acetylgalactosamine or β-1,4-linked galactose residues [9]. Galactosyl-residues are abundant on the surface of most cell types, thus the majority of eukaryotic cells are sensitive to ricin. Moreover, this toxin can also bind the mannose-type glycans on cells that carry those receptors i.e., macrophages or rat liver endothelial cells [45]. This is due to the fact that both the A- and B-chains of ricin are glycoproteins that contain mannose-rich oligosaccharides. However, in contrast to galactosyl-residues, most cell types do not express mannose receptors and consequently ricin in such cells can be internalized exclusively by galactosyl moieties. It is assumed that 10^6–10^8 ricin molecules can be bound to the cell surface [46]. Human cervical cancer cells (HeLa) for example contain 3×10^7 ricin-binding sites per cell [47], although not all of these sites are involved in toxin uptake.

After binding of ricin to the cell-surface receptors, the holotoxin is transported into the cell by endocytosis. It has been demonstrated that ricin is able to employ different endocytic mechanisms, which are believed to be mainly connected with the fact that it can recognize and bind to a great variety of cell surface components. It is even suggested that ricin can utilize all types of endocytic mechanisms that operate in a given cell [48]. First studies of ricin uptake demonstrated that it can be internalized by clathrin-dependent endocytosis in Vero cells (African green monkey kidney cells) [49]. However, further experiments showed that the inhibition of clathrin-coated pits formation in Hep-2 cells (derivative of HeLa) did not change ricin cytotoxicity [50] and that generally it can be taken into cells from non-coated parts of the cell surface membrane [51]. Additional evidence that ricin can be endocytosed by clathrin-independent mechanisms comes from experiments in which a dominant negative mutant of dynamin, a protein that is required for clathrin-related endocytosis, was used. The overexpression of the dynamin defective in GTP binding and hydrolysis in COS-7y cells (derived from kidney tissue of the African green monkey) did not alter cell sensitivity to ricin [52,53]. However, it should be noted that the expression of the same mutant dynamin (K44A) in HeLa cells inhibited ricin toxicity (see below). Moreover, it has been demonstrated that cholesterol plays an important role in endocytosis by clathrin-coated pits [54]. However, endocytosis of ricin was not changed significantly after cholesterol extraction from the membrane [54]. Effective extraction of cholesterol also disrupts other structures called caveolae that can be involved in clathrin-independent endocytosis [54,55], but independent experiments showed that caveolae were not necessary for ricin endocytosis [52]. Importantly, cholesterol is required for other endocytic mechanisms as well [56]. Thus, endocytic uptake of ricin can be both clathrin- and caveolae-independent [57]. Binding and endocytosis of ricin can be related to specific features of this toxin. Mutation that changes the secondary structure of its A-chain into a more helical structure (Figure 2A) has no influence on binding or endocytosis of modified holotoxin [58]. However, mutations that modify hydrophobicity of RTA (Figure 2B,C) significantly affect the binding of the altered holotoxins to the cell membrane [59]. A GFP-based reporter assay has very recently been applied in order to identify cellular components required for RTA intracellular transport in yeast [60]. The data indicated that proteins Syn8p, Sso1p and Snc1p influence ricin A-chain trafficking [60] (Figure 3). Syn8p and Snc1p form a complex that plays a role in protein uptake from the plasma membrane to endosomes [61]. Sso1p is a plasma membrane t-SNARE protein that together with v-SNARE Snc2p are required for the fusion of Golgi-derived vesicles with the plasma membrane [62]. It is unknown whether mammalian homologues of these proteins are also involved in ricin endocytosis.

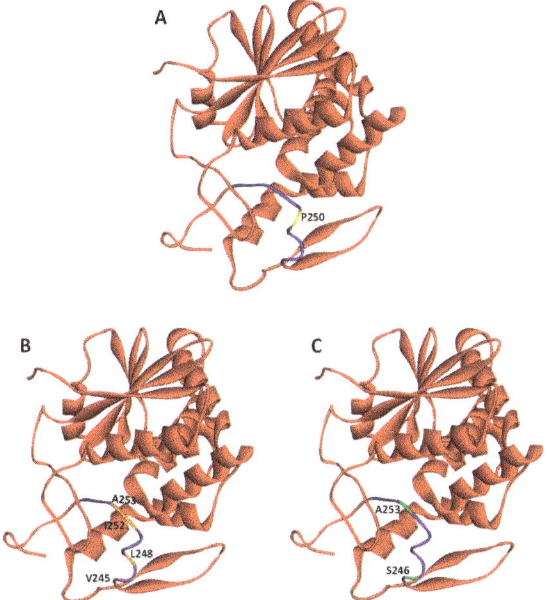

Figure 2. Crystal structures of ricin A-chain. PDB protein data bank (code 2AA1). Whole ricin A-chain (RTA) structures are marked in red. The hydrophobic region (Val245 to Val256) is indicated in purple. Several residues have been selected within the hydrophobic region and modified to obtain RTA with changed secondary structure or changed hydrophobicity. These residues are: (**A**) P250 (marked in yellow) has been modified (P250A) to produce RTA with a changed secondary structure; (**B**) V245, L248, I252, and A253 (marked in orange) have been modified (V245S, L248N, I252N, A253S) to obtain RTA with decreased hydrophobicity (RTA DHF); (**C**) S246 and A253 (marked in green) have been changed (S246V, A253V) to obtain RTA with increased hydrophobicity (RTA IHF).

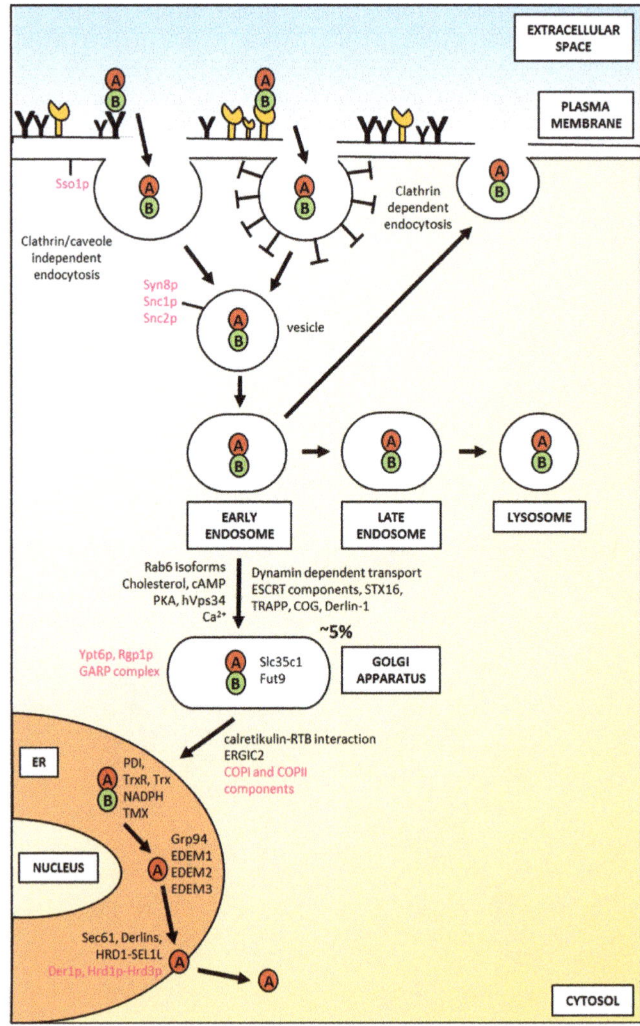

Figure 3. An overview of factors involved in the intracellular transport of ricin in mammalian and yeast cells. Yeast proteins are shown in magenta; mammalian proteins are shown in black. For a detailed description, please see the text.

2.2. Intracellular Routing of Ricin to the Endoplasmic Reticulum

After endocytosis ricin traffics to early endosomes (Figure 3). It has been demonstrated, that at this stage, the majority of toxin can be transported back to the cell surface in an apparently intact form [63]. In early endosomes, ricin also starts to be degraded [64]. This proteolysis is continued after its predominant transport to late endosomes and lysosomes [65] (Figure 3). Interestingly, efficient transport to lysosomes depends on how ricin is internalized. Significantly more toxin was transferred from endosomes to lysosomes upon internalization by mannose receptors in comparison to uptake by galactosyl-residues [45]. It was suggested that the different stabilities of ricin at endosomal low pH may be characteristic for these two binding mechanisms. This would explain the observed differences in the transport from endosomes to lysosomes between the two internalization pathways. In addition, as already mentioned in this review, modified ricin with a changed secondary structure that carries a point mutation in the hydrophobic region of the C-terminal A-chain (P250A) (Figure 2A) is more extensively degraded in endosomes/lysosomes than the wild-type toxin [57]. Both cathepsin B and cathepsin D were found to be involved in increased P250A ricin degradation. It cannot be excluded that P250 holotoxin has a somehow altered conformation in comparison to the wild-type toxin. This altered conformation might influence its endosomal/lysosomal degradation. It was visualized by electron microscopy that only about 5% of endocytosed ricin is transported to the trans-Golgi network (TGN) [66,67]. This transport proceeds directly from early endosomes, excluding the late endosomes-trans-Golgi pathway (Figure 3) [68]. From the Golgi complex, ricin is moved to the endoplasmic reticulum (ER). Ricin transport to the Golgi network and to the ER were confirmed experimentally [66,67,69]. Both pathways, endosome to Golgi and retrograde movement from the Golgi to the ER can engage complicated mechanisms of ricin intracellular transport. Brefeldin A is an important drug in these studies, because it disturbs the Golgi complex [70]. In cells treated with this compound, ricin transport was completely inhibited, and cell intoxication was blocked [71,72]. Interestingly, this was observed only in cells where the Golgi complex was sensitive to brefeldin A (BFA) [72].

In cells resistant to BFA, the polarized kidney epithelial cell line MDCK, BFA sensitized cells to ricin [73]. This drug does not change Golgi morphology in these cells and thus, it would be possible that increased transport to the Golgi apparatus is responsible for higher toxicity of ricin in MDCK cells. However, this did not seem to be the case. Moreover, it was demonstrated that BFA affects endosomal structures in those cells [73].

Ricin transport to the Golgi was also demonstrated by adding a short amino acid stretch, a tyrosine sulfation site that comes from rat cholecystokinin precursor. This extra stretch was added to the C-terminal end of the RTA, creating the ricin-A-sulf-1 [69]. When such a modified toxin was delivered to the Golgi complex, sulfotransferases that are specific enzymes for this compartment [74], catalyse the addition of sulfate to ricin. In cells that are preincubated with radioactive sulfate ($Na_2{}^{35}SO_4$,), ricin becomes labelled, indicating transport of this toxin to the Golgi apparatus. Although, dynamin is not required for ricin endocytosis [52,53], at least, it seems that transport from endosomes to the Golgi network is dynamin-dependent (Figure 3) in HeLa cells [75]. It was demonstrated that overexpression of mutant dynamin in HeLa cells did not affect ricin degradation but transport of endocytosed ricin to the Golgi and total ricin toxicity were strongly inhibited [75]. Since ricin transport after endocytosis is based on vesicles, it can be predicted that intracellular traffic of this toxin would depend on some Rab proteins, GTPases that regulate many steps of vesicular transport [76]. Moreover, knowledge about the involvement or lack of association of particular Rabs in ricin transport defines, at least partially, specific mechanisms of its intracellular routing. It seems to be the case for the Rab9-dependent pathway, that delivers proteins from late endosomes to the TGN [77,78]. Overexpression of a dominant-negative mutant of Rab9, (Rab9S21N), failed to protect cells from ricin intoxication and did not prevent sulfation of modified ricin-A-sulf-1 [79]. In addition, ricin transport to the TGN is independent of Rab-11, clathrin [79], as well as Rab7 [48]. Rab-9-independent transport of ricin supports the notion that this toxin may be transported to the Golgi complex directly from early endosomes. It has been demonstrated that isoforms of Rab6, Rab6A and Rab6A' are important in early endosomes-TGN transport, with a special role of Rab6A' in this pathway [80] (Figure 3). Both isoforms are expressed at similar levels in all cells. They differ in three amino acids located near one of their GTP-binding domains [81]. It was demonstrated that the ricin-related Shiga toxin B-chain utilizes direct transport from early endosomes being dependent on Rab6A' [82]. The hypothesis that ricin may use a similar, early endosomes-TGN pathway was confirmed by experiments demonstrating that its transport to the TNG is regulated by Rab6A and Rab6A' [83]. Moreover, ricin transport between endosomes and the Golgi network is regulated by changes in the cholesterol level. Ricin can be endocytosed by cells depleted of cholesterol [54], but transport to the TGN was strongly inhibited when the cholesterol was reduced [84]. Interestingly, ricin delivery to the Golgi complex is also calcium-dependent [85] (Figure 3). Calcium regulates many steps in intracellular trafficking, including intra Golgi transport [86]. It was observed that in cells treated with thapsigargin, which specifically inhibits the ER Ca^{2+}-ATPase, ricin transport to the Golgi was increased [85]. Moreover, genome-wide RNAi screens revealed human endosome-related genes such as the ESCRT component VPS2 (CHMP2A), Rab11FIP, and Rab5c as important for ricin toxicity [84]. Ricin transport also depends on the GARP complex, the SNARE Syntaxin16 (STX16), and the Golgi-related complexes TRAPP and COG [84] (Figure 3). However, VPS35, a central component of the Retromer that is required for the formation of transport carriers at the endosomes, is dispensable for ricin endosome-TGN transport. Mentioned already above, Rab6A and Rab6A' are human homologues of yeast Ypt6p [87]. Fluorescence-based reporter assay demonstrated that Ypt6p and its regulator Rgp1p are involved in RTA transport in yeast. Moreover, some GARP complex components (Vps51p and Vps54p) and Sft2p that enables fusion of endosome-derived vesicles with the Golgi are required for ricin traffic into cells [59]. Interestingly, Tlg2p (STX16 is a mammalian homologue of this protein) is not important for endosome-to-Golgi transport of RTA. In addition to the results described above, another simple reporter method based on the detection of changes in fluorescence emissions has been described and recently demonstrated to be useful for the identification of host cell proteins involved in intracellular RTA transport [88]. Endosome-to-Golgi transport of ricin is also regulated by cAMP signal transduction in the cell. It has been demonstrated that transport of ricin from endosomes to the Golgi network and further to the ER is controlled by the Golgi-associated regulatory subunit of protein kinase A (PKA) type II alpha isozyme in lymphocytes [89] (Figure 3). Moreover, transport of ricin to the Golgi is facilitated in human cells by hVps34 [90] (Figure 3), the only identified kinase that phosphorylates phosphatidylinositol (PI) in position 3 to produce PI(3)P. PI(3)P is needed for the vesicular localization of sorting nexins SNX2 and SNX4. All these events are required for ricin transport from endosomes to the Golgi complex [90].

Ricin transport to the Golgi complex can be observed by electron or immunofluorescence microscopy, whereas delivery of this toxin to the ER has never been directly visualized [91]. However, it was possible to use a genetically changed RTA containing two modifications: C-terminal Golgi-specific site for tyrosine sulfation (described above: RTA-sulf1) and three partly overlapping N-glycosylation sites (ricin-A-sulf-2) that flag an ER asparagine modification with carbohydrates [69]. It appeared that ricin-A-sulf-2 became sulfated in the TGN but also core glycosylated, indicating retrograde transport to the ER. Ricin itself does not contain an ER-targeting signal or KDEL retention sequence that would allow its interaction with KDEL receptors of the target cell and mediate toxin transport from the Golgi to the ER via coatomer protein I (COPI)-coated vesicles. However, RTB binds to resident luminal ER protein, calreticulin, a KDEL-tagged protein which indirectly allows for toxin transport to the ER [92]. Thus, calreticulin can operate as a retrograde transporter for ricin movement from the Golgi to the ER (Figure 3). However, calreticulin-deficient cells remained sensitive to this toxin, indicating that transport based on calreticulin-RTB interactions does not seem to be the main pathway for ricin traffic to the ER [92]. It was suggested that since ricin can bind glycolipids, some fraction of ricin may utilize lipid-sorting signals [93]. The second Golgi-to-ER transport pathway that was considered to be used by ricin is COPI-independent but Rab6-dependent [94]. However, it was demonstrated that the expression of the GDP-restricted mutant of Rab6A (Rab6A-T27N) did not alter ricin toxicity, suggesting that ricin is transported to the ER by a pathway that does not involve Rab6A [93]. Moreover, ricin was still toxic to cells when Rab6A and COPI were simultaneously inhibited. Thus, it was concluded that ricin can circumvent the Golgi apparatus. This hypothesis was confirmed in experiments in which the COPI protein complex was depleted of its subunit, epsilon-COP [94]. Cells were transfected with a modified ε-COP bearing a temperature-sensitive mutation. At the nonpermissive temperature, this protein become degraded, and the Golgi apparatus is changed morphologically. In such conditions, ricin could still be transported to the ER, even in the presence of brefeldin A which inhibits the binding of COPI to membranes and causes disassembly of the Golgi [95]. These results strongly suggest that ricin can bypass the Golgi stack on its way to the ER. However, it should be noted that such a pathway may be induced in the cells due to the changes enforced on these cells. Going back to the considerations about the classical Golgi-to-ER transport, it has been demonstrated by genome-wide screen that the ER-Golgi intermediate compartment protein 2 (ERGIC2) may be an important regulator of ricin transport to the ER [87]. siRNA-mediated downregulation of ERGIC2 protects cells against high doses of ricin. Consistent with these results, yeast homologue of ERGIC2 (Erv41p) and its complex partner Erv46p (mammalian ERGIC3) participates in RTA transport to the ER [59]. Both Erv46p and Erv41p are components of COPII vesicles, they form an active complex that traffics between the ER and the Golgi, being important for membrane fusion in ER/Golgi transport [96]. Additionally, the yeast reporter assay identified regulators of the ADP ribosylation factor (Arf) GTPases, Glo3p and Gea1p in ricin traffic in the cell. Arf initiates the budding of COPI-coated vesicles [97]. Two components of COPI, Sec22p and Rer1p are also important for ricin transport to the ER [96]. Rer1p is located at the Golgi membrane and operates as a retrieval receptor sending membrane proteins back to the ER [98]. Sec22p is a t-SNARE protein [99] that continuously traffics between the Golgi and the ER being involved in both anterograde and retrograde transport. Interestingly, it has been demonstrated that the mammalian homologue of Sec22p, Sec22B, is also important for ricin toxicity [87]. Thus, the components of both COPI and COPII seem to be important for ricin transport to the ER in yeast (Figure 3). Moreover, it has been suggested that COPII- and COPI-dependent ricin cycling between the ER and the Golgi is necessary for ricin dislocation to the cytosol and subsequent cytotoxicity [53]. It has been demonstrated very recently that GDP-fucose transporter residing in the Golgi, Slc35c1, and Fut9, a Golgi α1,3-fucosyltransferase, are both involved in fucosylation and are crucial for ricin toxicity [100] (Figure 3).

It should be noted that ricin A-chain can be directly delivered to the lumen of the ER by expressing a recombinant RTA version with an N-terminal signal peptide. The signal sequence is removed during RTA entry to the ER. Such an experimental approach appeared to be useful in yeasts [101,102], mammalian cells [103,104] and plants [105]. This procedure enables analyzing the events that happen after entry of the toxin into the ER, studying RTA interactions with ER-specific proteins, toxin transport from the ER to the cytosol and mechanisms of its intoxication. Directed transport to the ER is useful in yeast since yeast are deprived of galactosylated cell surface receptors [106] that are able to bind to the ricin B-chain. Thus, externally added ricin does not intoxicate yeast cells. Moreover, in mammalian cells, RTB was directed to the ER to study the formation of the disulfide bond between A- and B-chains in the ER [103]. In plants, preproricin is initially synthesized. It is composed of a single polypeptide chain of the RTA and RTB [107]. The first 35 amino acid residues of preproricin contain a 26 residue N-terminal signal sequence and a 9 residue propeptide [17]. The N-terminal signal sequence directs the transport of the nascent polypeptide across the ER membrane into the ER lumen; the 9 residue propeptide is removed after proricin transport to the vacuole [17]. Using the mature RTA (with 35 amino acid residues of preproricin) in tobacco leaf cells, it was demonstrated that a significant fraction of the newly synthesized ricin is retrotranslocated from the ER to the cytosol for degradation [105].

2.3. Ricin Translocation to the Cytosol

It is strongly believed that only active A-chain of ricin is translocated to the cytosol [53,69,104,108–110]; however, some suggestions that the whole holotoxin can be transported out of the ER have also appeared [111]. Anyway, it has been demonstrated that RTA transport to the cytosol is proceeded by reduction of the internal disulfide bond that connects the ricin A- and B-chains [103]. This reduction is catalyzed by the protein disulfide isomerase (PDI) [112], the main ER foldase that is responsible for the formation, cleavage and isomerisation of disulfide bridges [113]. It has been demonstrated that PDI interacts with the ricin B-chain and can both reduce and form the disulfide bond between ricin subunits [103]. However, it seems that the disulfide reductase activity of PDI needs to be enhanced by thioredoxin reductase (TrxR) [114] (Figure 3). In eukaryotes, two thiol-disulfide exchange systems exist: The thioredoxin system that contains thioredoxin and thioredoxin reductase [115] and the glutaredoxin system that includes glutaredoxin and glutathione reductase [116]. They catalyze fast and reversible reactions between cysteines in their active site and cysteines of their disulfide substrates using NADPH and reduced glutathione (GSH) as a source of reducing equivalents, respectively. In the case of ricin, it was demonstrated that PDI, TrxR and thioredoxin (Trx) used separately were unable to directly reduce ricin holotoxin [114]. However, PDI and Trx in the presence of TrxR and NADPH could release RTA from ricin holotoxin in vitro. PDI functioned only after pre-incubation with TrxR. The reductive activation of ricin was more efficient in the presence of glutathione [114]. Disulfide bond reduction in ricin holotoxin enables liberation of RTA, but it is also suggested that it serves to activate the catalytic activity of ricin A-chain [117]. It was demonstrated that recombinant proricin activity was dependent on its release from the mature ricin that was generated from proricin. Moreover, it was shown that another member of the PDI family, TMX, a transmembrane thioredoxin-related protein, reduces disulfide bridges in ricin holotoxin [118]. TMX can activate RTA by promoting interactions between ricin A-chain and ER proteins that facilitate transport of RTA to the cytosol prior to subsequent cell intoxication [118]. Increased ability of ricin to intoxicate cells after reductive release of RTA from holotoxin may result, at least partially, from the fact that liberated ricin A-chain can be unfolded to a certain extent in the ER. Such unfolding is probably required for RTA retrotranslocation to the cytosol through a relatively narrow ER channel. In support of this hypothesis, it was demonstrated that the introduction of an intrachain disulfide bond into the ricin A-chain significantly decreased the cytotoxicity of modified toxin [119]. This was explained by a constraint in the unfolding of RTA. Moreover, it was observed that a native A-chain is quite unstable at pH 7.0 [120]. Partially-unfolded RTA was sensitive to protease digestion and disrupted tertiary structure [120]. Thus, it was considered that ricin A-chain should be unfolded in the lumen of the ER, where it is recognized by several ER chaperones in a way similar to misfolded proteins.

Proteins that fail to become properly folded are recognized by specific ER factors that promote their transport to the cytosol for proteasomal degradation (for review see for example Refs. [121–124]). This process is called ER-associated degradation (ERAD) [122,125]. It is believed that the ricin A-chain, similarly to other A subunits of particular toxins, utilizes ERAD in its transport from the ER to the cytosol (for review see for example Ref. [110]). However, the main difference between RTA and typical ERAD substrates is that ricin A-chain avoids effective degradation by proteasome, being instead activated in the cytosol to exert its cytotoxic effect. Still, little is known about specific ERAD factors that facilitate RTA transport out of the ER. It has been demonstrated that both the A- and B-chain of ricin interact with one of the main Hsp70 chaperone family proteins, Bip (Grp78) [126]. Despite these interactions, overproduction of BiP significantly decreased RTA transport out of the ER and protected cells against this toxin. It cannot be excluded that ricin interacts with BiP that is already engaged in a bigger protein complex that forms in the ER and inhibits ricin transport to the cytosol. It has been demonstrated very recently that BiP can form a direct complex with Grp94 (Hsp90 chaperone protein) in the absence of a substrate [127]. Moreover, this interaction is nucleotide-specific. BiP and Grp94 more efficiently interact with each other at high ADP concentrations and possess lower affinity to interaction at high ATP concentrations. It has been demonstrated that inactivation of Grp94 by a specific inhibitor protects cells against ricin [128]. Except for classical chaperones (Hsps: 40, 70, 90 and 100), the ER possesses a unique class of carbohydrate-dependent lectins that recognize different ERAD substrates both in a glycan-dependent and independent manner [122,123,129]. The most known members of these lectin chaperones are calnexin/calreticulin [122,130] and the EDEM family [131–134]. Ricin B-chain interaction with calreticulin facilitates Golgi-ER transport (see above Ref. [92]), but it is not proven that this chaperone is directly involved in ricin transport to the cytosol. This is opposite to the role of EDEM1, EDEM2 and EDEM3. It has been demonstrated that RTA interacts with all EDEMs [135–138] (Figure 3). However, the mechanisms of their action during ricin translocation out of the ER are not the same. EDEM1 probably has a higher affinity for typical misfolded proteins than for ricin [135]. Thus, it can promote its transport to the cytosol only when ER translocons are not intensively occupied by ERAD substrates. EDEM2 directly facilities ricin transport to the cytosol, which induces higher cell sensitivity to this toxin [136]; whereas overproduction of EDEM3 is not relevant for RTA translocation to the cytosol [138]. Interestingly, ricin A-chain interaction with EDEM1 and EDEM2 and consequently its translocation to the cytosol and overall cytotoxicity is related to the appropriate structure and degree of hydrophobicity of RTA [57,58]. RTA P250A (with substitution of proline to alanine at amino acid position 250) of the highly hydrophobic C-terminal region (Val245 to Val256) has a changed secondary structure to a more helical one without alternations in RTA hydrophobicity [57] (Figure 2A). On the other hand, the substitutions V245S, L248N, I252N, A253S, S246V, and A253V in this region produce RTA with decreased (RTA DHF) (Figure 2B) and increased hydrophobicity (RTA IHF) (Figure 2C), respectively [58]. Both RTA P250A and RTA DHF show significant reduction in their ability for interactions with EDEM1 and EDEM2 [57,58]. Additionally, transport of RTA P250A to the cytosol and toxicity of this modified ricin were not dependent on EDEM1 and EDEM2 overproduction. These results demonstrate that for interactions between EDEM1, EDEM2 and RTA, appropriate structure and hydrophobicity of the substrate are important. RTA with a decreased amount of β-sheet structures that directly resulted from increased α-helicality [57] and RTA with very low hydrophobicity of the C-terminal region [58] exhibit reduced interactions with EDEM chaperone proteins. Interestingly, it was demonstrated that a conformational change of RTA is crucial for its binding to the surface of the ER membrane. At the physiologically relevant temperature of 37°C, RTA loses some of its helical content and rearranges the conformational structure in such a way that it exposes its C-terminal region to the membrane interior [139]. Such an insertion into the ER membrane might be necessary for RTA translocation to the cytosol. It can be concluded from these observations that an additional limiting step in RTA P250A retrotranslocation to the cytosol (apart from lack of EDEM1 and EDEM2 assistance) might result from its inability to be subjected to additional conformational changes allowing it to be stably inserted into the ER membrane. As already mentioned, EDEM proteins can recognize glycan residues present on their substrates but can also bind other structures, e.g., hydrophobic regions or unfolded motifs. This second option seems to be important for EDEMs–ricin interactions since recombinant ricin expressed in *E. coli* that was used in the experiments, lacks oligosaccharides that are normally added to ricin A-chain derived from plants [140]. On the other hand, it was demonstrated in *S. cerevisiae* that RTA glycosylation (that occurs on asparagines 10 and 236) promotes its transport from the ER to the cytosol and increases ricin cytotoxicity as block in RTA glycosylation impairs depurination of specific adenine in 28S rRNA [141]. Moreover, in the case of ricin, a glycan signal can stabilize this toxin [142]. This is opposite to typical misfolded proteins where specific oligosaccharide recognition becomes a signal for their degradation. Ricin stabilization was demonstrated by using GFP-tagged RTA containing a point mutation (E177Q) which attenuates its cytotoxicity (GFP-RTA E177Q). This toxin, engineered with a murine signal sequence for direct co-translational delivery into the ER of the host cell, was destabilized by inactivating genes required to generate and recognize the N-glycan residues [142].

In its transport from the ER to the cytosol, ricin A-chain definitely utilizes one specific type or different classes of the ER membrane translocation channels (Figure 3). Three main types of ER translocons have been identified so far: the Sec61 complex [143–147], the Derlin proteins [148–153] and several ER membrane multi-spanning ubiquitin ligases, including HRD1 (Hrd1p in yeast) [154–159]. The main function of the ubiquitin ligases is connected with polyubiquitination of polypeptides emerging in the cytosol prior to their transfer for the proteasomal degradation (for review see for example Ref. [160]). However, the ubiquitin ligases can also form a translocation channel being directly involved in ERAD substrates' transport to the cytosol [156,157]. The role of Sec61 and Derlin proteins in RTA transport out of the ER is not clear and unambiguous. It is still discussed whether these channels can be used by ricin A-chain as real translocons. It has been demonstrated that RTA can interact with Sec61α, the main component of the Sec61 complex in mammalian cells. This was shown by co-immunoprecipitation studies [108,135] and additionally demonstrated with isolated yeast ER-derived microsomes [101]. Moreover, the rate of RTA degradation was significantly decreased in yeast mutants defective in protein export via the Sec61p translocon [101]. On the other hand, genome-wide RNAi screens did not identify Sec61 as important in ricin toxicity [87]. In these experiments, Sec61 was effectively downregulated, as a mix of the single most potent siRNA against each gene of the Sec61 complex was used. In addition, gene silencing of *Sec61α* did not influence RTA transport from the ER to the cytosol in HEK293 (human embryonic kidney) cells [161].

However, it cannot be excluded that unchanged ricin A-chain translocation to the cytosol upon Sec61α downregulation might result from the existence of undefined compensatory mechanisms that direct ricin to other ER channels. In the case of Derlins, the majority of collected data indicate that these proteins are not crucial for RTA transport to the cytosol. The rate of RTA degradation was not decreased in yeast cells devoid of Der1p (mammalian Derlin-1) [101]. Cells stably transfected with dominant negative constructs of Derlin-1 and Derlin-2 treated with extrinsically added ricin [135], as well as dominant negative Derlin-1 transfected mammalian cells expressing an ER-localized RTA construct [104], did not exhibit altered ricin A-chain transport to the cytosol when compared to the control cells. Moreover, overproduction of Derlin-1 or Derlin-2 did not influence RTA dislocation to the cytosol [162]. In contrast, single Derlin protein downregulation as well as gene silencing of all three Derlins (*Derlin-1*, *Derlin-2* and *Derlin-3*) showed significant rescue against ricin intoxication [87]. The assumption that Derlins can play a specific role in ricin transport across the ER membrane was strengthened by an observation showing that two factors, UFD1L and NPLOC4, that bind to Derlins, are required for ricin intoxication [87]. However, considering the effect of Derlin-1 on overall ricin cytotoxicity, it should be noted that the ER-cytosol step is not the only one that contributes to this process, since it was demonstrated that Derlin-1 is necessary for an efficient retrograde transport of ricin from endosomes to the Golgi apparatus [163] (Figure 3). The role of HRD1 ubiquitin ligase and its cofactor SEL1L seems to be the most obvious in RTA translocation to the cytosol (Figure 3). It was shown that SEL1L is required for ricin A-chain transport to the cytosol and *SEL1L* knockdown protects cells from ricin [104]. Similarly in yeast, Hrd1 and its cofactor Hrd3p facilitate RTA transport out of the ER [164]. However, recently published results demonstrated that Hrd1p and Derl2 (mammalian Derlin-2) contribute to, but do not exert, an absolute requirement for ricin intoxication [165].

3. Cytotoxic Action of Ricin on Cells

3.1. Activation of Ricin A-Chain in the Cytosol Regulation of RTA Folding versus Degradation

After translocation to the cytosol, ricin must refold into its biologically active conformation to modify its cytosolic targets (Figure 4). It is considered that there are three general pathways by which ricin A-chain can obtain its catalytic, folded structure: binding to cytosolic chaperones [53,128,166], ribosome-mediated refolding [120], and interactions with ribosomal and proteasomal factors [166]. However, it should be noted that not all of the RTA molecules translocated to the cytosol can act as an active toxin. Ricin A-chain is partially degraded by the 26S proteasome and this degradation can be blocked by specific proteasome inhibitors [57,108,135,136,167]. It seems that in mammalian cells, the majority of RTA escape proteasomal degradation [108,135]. However, in yeasts, it was shown by pulse-chase experiments that only 20% of translocated toxin appeared to be completely stable, whereas the rest was degraded during the first hour of the chase [101]. Cell fractionation has shown that this stable RTA was present in the cytosol. Interestingly, yeast proteasomes discriminate between native and structurally defective forms of RTA [164]; native RTA can avoid proteasomal degradation. This phenomenon is explained by the fact that ricin A-chain contains only two lysine residues which generally do not become efficiently ubiquitinated during toxin transport to the cytosol [109,167,168]. The lack of RTA lysine ubiqitination is common for both yeast and mammalian cells despite the fact that requirements for the ubiquitin-dependent system, Cdc48p/p97 that extracts ricin A-chain out of the ER membrane differ between these two groups. In yeast, RTA transport to the cytosol is independent of Cdc48 [164], whereas in mammalian cells, the expression of a dominant negative mutant of p97 blocked ricin toxicity and increased the time required for RTA transport to the cytosol [169]. The second general mechanism that allows toxins to avoid proteasomal degradation might be connected with their ability to obtain its fully-folded structure relatively quickly after transport to the cytosol. It has been demonstrated that folded proteins do not become proteasomal substrates even if they possess many lysine residues on their surface [170]. However, in the case of ricin, it was shown that it cannot refold spontaneously after thermal denaturation in vivo [120]. At 37°C, ricin has a conformation similar to a molten globule, and it was impossible to obtain its native state by manipulation of the buffer conditions or by the addition of a stem-loop dodecaribonucleotide or deproteinized *E. coli* rRNA, both of which are substrates for ricin A-chain. Thus, RTA is considered to be a toxin with a slow refolding rate. This rate is much slower than for example the A subunit of the cholera toxin (CTA1) that displays lower detectable sensitivity to degradation by the proteasome when compared with ricin [171].

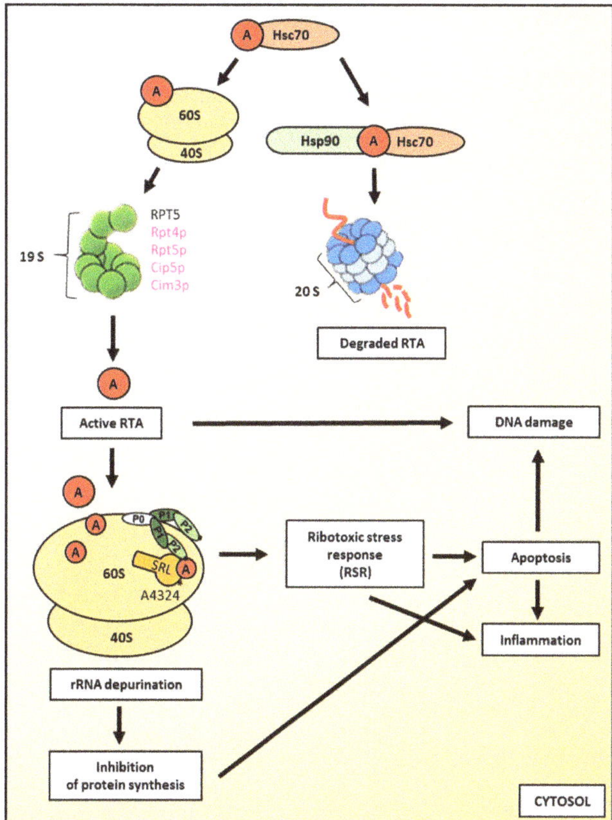

Figure 4. Cytotoxic action of ricin on cells. Binding to cytosolic chaperones, ribosome-mediated refolding and interactions with ribosomal and proteasomal factors are pathways by which ricin A-chain can obtain its active form. However, ricin A-chain is partially degraded by proteasomes. Active ricin A-chain is an *N*-glycosidase that removes a universally-conserved adenine at position 4324 from the α–sarcin-ricin loop (SRL) of the rRNA present in the large ribosomal subunit. The interaction of ricin A-chain with the large ribosomal subunit is facilitated by the ribosomal stalk structure, composed of P0, P1 and P2 proteins. The damage of the 28S rRNA by ricin leads to the inhibition of protein synthesis and triggers the ribotoxic stress response (RSR). Both pathways can induce apoptosis and further inflammation. Ricin A-chain can also directly induce DNA damage. Yeast proteins are shown in magenta; mammalian proteins are shown in black. 20S refers to the core particle, whereas 19S refers to the regulatory particle of the proteasome. For a detailed description, please see the text.

The interactions of ricin A-chain with ribosomal proteins and cytosolic chaperones are crucial in gaining its proper conformation. These interactions also facilitate RTA transport to the cytosol. It has been demonstrated that an ATPase subunit of the 19S proteasome cap in yeast, Rpt4p [164], and two other proteins of the cap, Cim3p and Cip5p [101], are important for ricin A-chain transport from the ER.

However, no obvious requirement for this transport was observed for the other Rpt subunits, Ubr1p or the proteasome core itself [164]. It should be noted that Rpt4p can cooperate with Cdc48p in the extraction of an endogenous substrate from the yeast ER [172]. Chaperone-like activity of ribosomes were demonstrated in experiments showing that partially-unfolded ricin A-chain was able to obtain full catalytic activity in the presence of salt-washed ribosomes [120]. The ATPase subunit RPT5 of the 19S proteasome cap prevents aggregation of denatured RTA and enhances the recovery of catalytic activity of ricin A-chain in vitro [166]. In addition, it was shown in vivo that Rpt5p is required for maximum toxicity of RTA dislocated from the yeast ER. Interestingly, the anti-aggregation properties of the 26S proteasome are independent of its proteolytic activities [166]. Cytosolic chaperones protect and activate ricin A-chain but they also have a much broader spectrum of action, as they regulate the general fate of RTA dislocated from the ER [128]. The mechanisms of cytosolic chaperones activity include ricin A-chain binding to Hsc70 (cytosolic member of Hsp70 family). It was shown that Hsc70 prevents aggregation of the heat-inactivated toxin and can recover its catalytic activity. Inhibition of cytosolic Hsc70 protected HeLa cells from ricin [128]. However, the concentration of Hsc70 co-chaperones may regulate the amount of RTA that can gain the catalytic activity and the fraction that will be degraded. Interaction of the RTA-Hsc70 complex with BAG-2 and Hip induces RTA activity in vitro, promoting the sensitivity of cells to ricin. On the other hand, co-chaperones BAG1 and CHIP facilitate ricin A-chain destabilisation. CHIP is an E3 ubiquitin ligase that interacts not only with Hsp70 but also with another type of cytosolic chaperone, Hsp90. Moreover, another dual co-chaperone, Hop, is an Hsp70-Hsp90 organizing protein. It reversibly links Hsp70 and Hsp90 by recruiting Hsp90 to the existing Hsp70-substate protein complex. This promotes transfer of the substrate from Hsc70 (Hsp70) to Hsp90 [173]. Overexpression of Hop decreases sensitivity to ricin, suggesting that sequential interaction of RTA with Hsc70 and Hsp90 directs the toxin to inactivation and destabilisation [128]. Consistent with this hypothesis, inhibition of Hsp90 sensitized cells to ricin [128]. RTA inactivation may be mediated by an ubiquitination process. This mechanism is not fully elucidated, but it was suggested that a low amount of RTA can be ubiquitinated in the cytosol. In yeast, ubiquitination occurs via an unknown E3 ligase [164]. It has been demonstrated that RTA is not ubiquitinated by Hrd1p during dislocation [164], but introduction of additional lysyl content into RTA reduces its cytotoxicity by increasing ubiquitin-mediated proteasomal degradation [168]. In yeast, no significant changes in the growth rate were observed in cells lacking individual Hsp40, Hsp70 and Hsp90 family members or the Hsp70 and Hsp90 co-chaperones [164].

3.2. Ricin A-Chain Action on Ribosomes

Ricin A-chain is an *N*-glycosidase that removes a universally-conserved adenine at position 4324 in mammalian cells (Figure 4) and A3027 in yeast from the α–sarcin-ricin loop (SRL) of the rRNA present in the large ribosomal subunit ([5,8,10,11,174] and see Introduction for details). This disables the binding of specific elongation factors to the ribosome and inhibits protein synthesis [9,16,17]. It is known that ricin A-chain influences the structure of the ribosomal RNA. It alters the dynamic flexibility of the GTPase activating centre of the ribosome, which part is SRL. These conformational changes disturb the transition between the pre-and post-translocational states of the elongation cycle [175]. As earlier mentioned in this article, ricin does not depurinate *E. coli* ribosomes. However, its inability to exert a cytotoxic effect on prokaryotic ribosomes does not result from the fact that RTA is incapable to act on the 23S rRNA from the large bacterial ribosomal subunit. In fact, ricin A-chain depurinates the sarcin-ricin loop of naked 23S rRNA [176], suggesting that for the proper catalytic activity of ricin, the whole ribosome and particularly ribosomal proteins are crucial [177]. It has been demonstrated that the ribosomal stalk structure facilitates the interaction of ricin A-chain with the large ribosomal subunit in eukaryotes [177–179] (Figure 4). The human ribosomal stalk structure is composed of three types of phosphoproteins, P0, P1 and P2. They are assembled into a pentameric protein complex comprised of a single P0 protein bound by two heterodimers of P1 and P2 proteins [180,181]. It has been shown recently that P1–P2 proteins represent a primary binding site for RTA [182], with a more critical role for the P1B–P2A dimer [183]. Interestingly, the stalk structure differs significantly between prokaryotic and eukaryotic ribosomes. In prokaryotic cells, the stalk proteins exhibit very little sequence homology to eukaryotic counterparts. Thus, it was suggested that they are solely functionally analogous to eukaryotic proteins [184]. This important observation significantly contributes to our understanding of species specificity for ricin and ultimately defines the sensibility of the SRL for depurination by this toxin. The ribosomal stalk is a dynamic structure located in proximity to the SRL. It was suggested that ricin A-chain binding to the stalk proteins may allow RTA to be placed directly by the sarcin-ricin domain, which facilitates more efficient detection of the rRNA substrate by RTA [177]. Based on the analysis of interactions between RTA and the stalk proteins, a two-step binding model was proposed with the first step characterized by a slow association and dissociation rate, and a second step with much faster rates of interactions [185,186]. Within this model, the binding of RTA to the stalk was then more specifically divided into four phases. In the first phase of binding, ricin A-chain is concentrated on the surface of the ribosome and directed to the stalk [185,186]. This step is based on slow and nonspecific electrostatic interactions. In the next phase, RTA interacts with the stalk through more specific and stronger electrostatic interactions, which are saturable. Step three represents the delivery of RTA to the SRL [185,186]. This transfer is mediated by the C-terminal domain (CTD) of the stalk proteins. Their involvement in this process decreases the possibility of RTA dissociation out of the ribosome and leads to rapid recruitment of the toxin by SRL. Finally, in the fourth phase, ricin A-chain specifically depurinates the α-sarcin-ricin loop [185,186]. Recently published results demonstrated that the C-terminal domain of P1 is the main docking site for RTA as deletion of P1 CDT but not P2 CDT decreased the affinity of the stalk structure for ricin A-chain [182]. However, studies of another group have demonstrated that RTA mainly recognizes the highly conserved C-terminal tail of P2 (residues 106–115) in which two residues, Leu and Phe, are critical for the interaction with RTA [187]. These residues might also be important for RTA binding to other P proteins. Moreover, the crystal structure of RTA with P2 protein shows that GFGLFD motif of the C-terminal P2 is inserted into a hydrophobic pocket of RTA, suggesting that the flexibility of the P2 peptide interaction with ricin A-chain is based on hydrophobic rather than electrostatic interactions [188]. It was shown before that seven arginine residues located at the RTA/RTB interface are involved in ricin A-chain interaction with the ribosome [189,190]. In the holotoxin structure, each arginine residue is covered by RTB, completely or only partially [191]. Nevertheless, this causes the ribosome binding site on RTA to be blocked by RTB, thereby disabling ricin holotoxin to depurinate ribosomal rRNA. Thus, holotoxin RTA can become catalytically active only after its release from RTB which results in revealing the ribosome binding site [192,193]. Interestingly, RTA with a modified ribosome binding site is less toxic than a variant with lower catalytic activity but unchanged ribosome binding activity [194]. It has been demonstrated that by introducing R189A/R234A and R193A/R235A double mutations, ricin A-chain binding to the stalk stimulates ribosome depurination by orienting the active site of RTA toward the SRL, thereby allowing docking of the target adenine into the active site [192]. Recently published results also showed that Arg235 of ricin A-chain serves as a main interacting residue with ribosomes and cooperates with nearby arginines to allow RTA to interact with the stalk with fast kinetics in order to achieve the binding specificity necessary for SRL depurination [191]. It was initially proposed, based on a study with analytical ultracentrifugation, that RTA interacts with the 60S ribosomal subunit with a molar stoichiometry of 1:1 [195]. However, a model involving conformational changes is currently preferred [196]. In this model, after binding of RTA to the ribosome, conformational rearrangements of both RTA and ribosomes occur, allowing the formation of a high affinity complex. These conformational changes directly influence the catalytic activity of ricin A-chain. It has recently been shown that the flexibility of the α-helix (residues 99–106) of RTA is connected with the regulation of the depurination activity by ricin A-chain, which directly influences the rate of protein synthesis inhibition [197]. Moreover, it was proposed that the flexibility of the α-helix could affect the side chain orientation of Glu-177, which is critical for the depurination activity of ricin [198,199]. Interestingly, ricin A-chain is not able to bind to the isolated 40S ribosomal subunit, however, its rRNA and/or ribosomal proteins may promote optimal and stable interactions of ricin with the whole ribosome, since binding of RTA to 80S ribosomes was approximately 3.5-fold stronger than binding to the isolated 60S subunits [177,195]. The interaction of ricin A-chain with the ribosomal stalk structure has been evaluated as a potential drug target [200]. Several peptides (3 to 11 amino acids in length) corresponding to the C-terminal end of P proteins were examined in their ability to interact with RTA and block its catalytic activity. It appeared that a four amino acid peptide is the shortest one that can inhibit depurination activity of RTA by preventing toxin binding to the ribosome [200].

It has been demonstrated that one molecule of ricin A-chain is able to inactivate over 1500 ribosomes per minute in a cell-free preparation of ribosomes [18,19]. The multiplexed digital droplet (ddPCR) assay showed that depurination events in lung cell cultures can be detected as early as 1 h after ricin treatment (1 nM) and within 9 h of exposure the maximum ribosomal damage of 70% was reached [201]. This effect was sustained for at least 24 h post-exposure. However, it should be noted that depurination in cell-based systems would be expected to occur at a lower rate than in cell-free systems. In the cell, the whole A-B holotoxin is applied and there is a lag time that is required for toxin uptake and its intracellular transport until the rRNA substrate is reached. Depurination rates are difficult to compare between in vivo and in vitro experiments, moreover, the rates of depurination may differ between cell types [176,201,202]. Nevertheless, it can be assumed that ricin-induced depurination is a very rapid enzymatic process.

3.3. Mechanisms of Ricin-Induced Apoptosis

Despite the very effective ricin-mediated rRNA depurination event that results in the inhibition of protein synthesis, it cannot be stated that these processes by themselves lead to cell death [24,30–33,203]. It has been demonstrated that ricin can induce apoptosis, autophagy and release of cytokine inflammatory mediators [25–30,204] (Figure 4). These processes have been studied for over two decades with a growing number of results that indicate their significance for cell death observed. However, it is still unclear whether the inhibition of protein synthesis is sufficient to induce apoptosis in all cell lines and to what extent other factors are required for the induction of apoptosis [30,205].

Early reports describing that ricin is capable of inducing cell death by apoptosis came from in-vitro studies that correlated cellular morphological changes with apoptosis [26,206–208] and from in-vivo studies in which epithelial, endothelial and myeloid cells were used [204,209–211]. It was observed that cells treated with ricin exhibit chromatin condensation, membrane blebbing, rounding of the cells, formation of apoptotic-like bodies, and DNA fragmentation that are considered to be typical markers of apoptosis [26,27,206,207,209,210]. Moreover, it has been reported that intracellular targets of ricin are not limited only to the ribosomal RNA. Deproteinized (naked) RNA, synthetic oligoribonucleotides, nuclear and mitochondrial DNA, polyA, tRNA and viral nucleic acids were published to be depurinated by purified ricin and other type II RIPs [211]. These ricin activities were often classified as actions not directly connected with apoptosis. It has been reported that the early DNA damage observed in human endothelial cells HUVEC in parallel to the arrest of protein synthesis was not a consequence of ribosome inactivation or apoptosis but results from direct action of ricin on DNA [212] (Figure 4). Li and Pestka [213] suggested that ricin and other RIPs may induce rRNA damage, in addition to the classical way, also through increased expression and activation of host RNases. Moreover, ricin-mediated rRNA depurination might facilitate toxin interaction with one or both dsRNA-binding domains of a kinase associated with the ribosome, PKR (double-stranded RNA-activated protein kinase), thereby causing the activation of PKR [214]. PKR plays a role in interleukine-8 (IL-8) induction, which may trigger the ribotoxic stress response (RSR) (see below).

It is considered that the main mechanisms of ricin-dependent apoptosis are based on the activation of caspases [215–222], Bcl-2 family members [223–225], and stress-associated signaling pathways [102,213,215,219,226–238]. The mechanisms of ricin-induced cell death promoting pathways are also connected with direct and indirect action of ricin on DNA [212,239], ricin-mediated reactive oxygen species production [24,215,218,224,240–242], and ricin B-chain-induced apoptosis [33,243].

3.3.1. Ricin-Induced Activation of Caspases

It is assumed that apoptosis can be activated via two major pathways, extrinsic and intrinsic. The extrinsic or receptor-mediated mechanism includes ligation of the death receptors that stimulate the activation of the initiator caspase-8, which then triggers downstream events by direct stimulation of caspase-3 or cleavage of the protein Bid. In the intrinsic pathway, mitochondria function as the main operation center. Damage to mitochondria results in outer membrane depolarization and permeabilization (MOMP) which triggers the release of several proapoptotic factors, including cytochrome c. This leads to the activation of caspase-9, which then triggers effector caspase-3 [244,245]. Poly (ADP-ribose) polymerase (PARP) is one of the best described substrates for caspase-3 which can cleave 116kD PARP into 85 and 31kD fragments [246]. PARP cleavage is one of the most studied hallmarks of apoptosis. It is believed that mitochondria and the intrinsic pathway of apoptosis activation are critical in signaling for cell death in ricin-intoxicated cells. It has been reported that in ricin-treated cells, a loss in mitochondrial membrane potential, rapid release of cytochrome c, activation of caspase-9 and caspase-3, and DNA fragmentation were observed [215–222]. Caspases may be profoundly involved in the pathway, resulting in DNA fragmentation [220,247]. Moreover, ricin-dependent PARP cleavage has been observed in different cell lines including HeLa [218] and U937 cells [216,220]. Interestingly, studies performed in U937 cells might, at least partially, answer the question about correlation between different ricin-mediated death mechanisms in cells. In ricin-treated U937 cells, intracellular NAD(+) and ATP levels were decreased and this reduction was followed by ricin-mediated protein synthesis inhibition [220]. The PARP inhibitor, 3-aminobenzamide (3-ABA), blocked the depletion in NAD(+) and ATP levels. Significant PARP cleavage was observed more than 12 h after ricin addition, while DNA fragmentation reached a maximum level within 6 h of incubation [220]. Thus, it was concluded that the PARP cleavage is not an early apoptotic event associated with the induction of ricin-mediated apoptosis and that the pathway leading to cell lysis via PARP activation and NAD(+) depletion is independent of the pathway leading to DNA fragmentation. Moreover, it seems that human BAT3 (HLA-B-associated transcript 3, Scynthe) [248] is an important regulator of ricin-dependent caspase-3-mediated apoptosis [215,221]. It interacts with ricin A-chain, which was confirmed by co-immunoprecipitation and confocal microscopy studies. BAT3 possesses at its C-terminal end, a canonical caspase-3 cleavage site, thus being the substrate for this protease. As a result of this cleavage, a 131 amino acid C-terminal fragment of BAT3 (CTF-131) is generated. It was observed that ricin-mediated induction of cell apoptosis by caspase-3 activation resulted in BAT3 cleavage after 4 h treatment with ricin. On the other hand, ricin-induced apoptosis was significantly reduced in cells with a decreased level of BAT3. Importantly, CTF-131 but not BAT3 was responsible for the observed direct ricin-induced apoptotic morphological changes such as: cell rounding, nuclear condensation, and phosphatidylserine exposure [221]. It was also observed that internalized ricin co-localized with endogenous BAT3 in the nuclei of HeLa cells. It was suggested that caspase-3-mediated proteolysis of BAT3 may require caspase-3 translocation to the nucleus. Another important regulatory factor is the mitochondrial intramembrane protein AIF (apoptosis-inducing factor). BAT3 interacts with AIF in the cytosol [248], regulates its stability by inhibiting proteosomal degradation and also induces nuclear translocation of this factor. AIF may regulate caspase activation, however, after translocation to the nucleus, it is involved in chromatin condensation and DNA fragmentation [248].

As described above, the majority of experiments promote the idea that ricin initiates the intrinsic, mitochondrial-dependent pathway of apoptosis. However, some findings suggest that the extrinsic pathway might be also important in ricin-induced apoptosis. It has been demonstrated by TUNEL immunohistochemical staining, flow cytometry, and Western blotting that purified ricin A-chain was able to induce apoptosis in mouse embryonic fibroblast (NIH 3T3), by activation of caspase-8 and -3 but not caspase-9 [222].

3.3.2. Activation of Bcl-2 Family Members by Ricin

Bcl-2 (B-cell lymphoma protein-2) family members regulate the intrinsic pathway of apoptosis. This family of proteins consists of three subfamilies playing opposing functions in this process: proapoptotic BH3-only members (Bim, Bid, Puma, Noxa, Hrk, Bmf, and Bad), proapoptotic effector molecules (Bax and Bak), and antiapoptotic Bcl-2 family proteins (Bcl-2, Bcl-xL, Mcl1, A1, and Bcl-B) [244,245,249]. It has been shown that overexpression of Bcl-2 improves the growth of MCF-7 breast cancer cells treated with ricin by 10-fold [223]. However, ricin retained its ability to inhibit protein synthesis in those cells. It has also been demonstrated that the ricin-induced apoptosis of hepatoma cells, BEL7404, results from increased expression of Bak and decreased levels of Bcl-xl and Bax [224]. However, overexpression of Bcl-2 can protect BEL7404 against ricin. These results are in agreement with other observations, supporting the view that signaling through mitochondria can represent the main mechanism of ricin-induced apoptosis. It is possible that in cells with an elevated level of Bcl-2, ricin-induced cell death is inhibited through titrating the function of its pro-apoptotic homologues, such as Bax. Consistent with these results, in cells overexpressing Bcl-2, a lower level of ricin-induced caspase-3 activity and PARP cleavage were observed in comparison to control BEL7404 cells treated with ricin [224]. Interestingly, the use of a caspase-1-specific inhibitor also partially blocked ricin-induced apoptosis, implicating a role for caspase-1, and therefore possible involvement of the inflammasome and cytokine production in this process. Other studies showed that BER-40 cells (brefeldin A-resistant mutant cell line of Vero) were highly resistant to ricin-induced apoptosis as compared with their none-modified counterparts, parental Vero cells [225]. It was suggested that the function of mitochondria may be somehow altered in BER-40 since a lack of release of cytochrome c was observed in these cells. However, the number and structure of mitochondria were not changed in these cells. Also, the expression level of Bcl-2 (which is the regulatory protein involved in the release of cytochrome c), was the same in Vero and BER-40 cells treated with ricin. Thus, it was suggested that relatively early apoptotic signaling pathways prior to those that lead to the release of cytochrome c may be altered in BER-40 cells. However, there is a possibility that mitochondria-related factors such as Bax and Bcl-xl, or the regulation of Bcl-2 activity by phosphorylation and dephosphorylation, might also be involved in the apoptosis resistance phenotype in BER-40 cells [225].

3.3.3. Activation of Stress Associated Signaling Pathways by Ricin

It has been proposed that the damage of the 28S rRNA by ricin triggers a specific kinase-activated pathway termed the ribotoxic stress response (RSR) [226]. In this signaling pathway, stress-activated protein kinase SAPK/JNK1 (c-Jun N-terminal kinase) is activated together with its activator kinase SEK1/MKK4 [250]. It has been demonstrated that this activation does not result from the inhibition of protein synthesis, but is directly connected with signaling from the 28S rRNA affected by ricin. JNKs, p38 and extracellular-receptor kinases (ERKs) belong to the Ser/Thr kinases termed MAPK (mitogen-activated protein kinase) family [251]. It is known that ricin can activate not only JNK, but also ERK and p38 MAPK in RAW 264.7 macrophages, and this is necessary for further activation of a variety of proinflammatory mediators [227]. Interestingly, not only ricin holotoxin but also high concentrations of ricin A-chain were able to induce both the p38 and JNK MAP kinase signaling pathways; however, signaling through the JNK kinase appeared to be more important in inducing the apoptotic response by RTA in the nontransformed epithelial cell line, MAC-T cells [219]. Ricin treatment induces the expression of proinflammatory cytokines and chemokines such as TNF-α, interleukin (IL)-1, IL-6, and IL-8 [227–231]. It was demonstrated that macrophages and IL-1 signaling play a central role in the inflammatory process triggered by ricin [232]. Moreover, ricin is an activator of the NALP3 inflammasome, a scaffolding complex that mediates pro-IL-1β cleavage to active IL-1β by caspase-1 [233], (for review see also Ref. [234]). The proinflammatory response of ricin is believed to be initiated by phosphorylation of the kinase ZAK, a MAP3K, that is located upstream to the kinases p38 MAPK and JNK in a signal transduction pathway leading to proinflammatory gene expression [235]. It has been demonstrated that the JNK and p38 pathways regulate the expression of cytokines and downstream transcription factors in a different way [227]. The use of specific chemical inhibitors of the SAPK pathways in ricin-treated RAW 264.7 macrophages showed that suppression of the p38 pathway almost completely inhibited IL-1α and -β expression, while blocking the JNK pathway increased the expression of these cytokines. In contrast, inhibition of both pathways equally attenuated the ability of ricin to induce TNF-α gene expression [227]. Moreover, the role of p38 in ricin-induced expression of various proinflammatory genes was demonstrated [229], and chemical inhibition of the p38 pathway in the human monocyte/macrophage cell line 28SC blocked ricin-induced IL-8 secretion [228]. In addition, inhibition of the p38 MAPK pathway in ricin-treated RAW 264.7 macrophages attenuated both TNF-α secretion and apoptosis [236]. It was concluded that the ribotoxic stress response may trigger multiple signal transduction pathways through the activation of p38 MAP kinase. These results underline the major role of MAPK kinases and especially p38 in ricin-dependent regulation of apoptosis and in proinflammatory signals gene expression.

Besides MAPKs activation, ricin can also trigger the NF-κB pathway that is responsible for regulation of the expression of genes encoding inflammatory and pro-coagulant mediators [215,230]. It was suggested that the inhibition of protein synthesis by ricin may lead to the activation of NF-κB. Moreover, the activation of both the JNK and p38 MAPK pathway as well as NF-κB occurs independently. Inhibition of TNF-α in cultured primary human airway epithelial cells did not prevent ricin-induced activation of NF-κB [230]. However, inhibition of NF-κB resulted in the release of cytochrome c from the mitochondria [252] and JNK1 kinase activation [252,253], suggesting an anti-apoptotic function of this transcription factor. The regulation of survival and apoptotic signals triggered by MAPKs and NF-κB in response to ricin remains to be determined.

The second stress-associated signaling pathway affected by ricin is the unfolded protein response (UPR), which is related to the ER stress. This signaling pathway can be characterized as a cell reaction to the accumulation of unfolded or misfolded proteins in the lumen of the ER [122,124,159,254]. The UPR is regulated by three ER transmembrane receptors: the RNA-dependent protein kinase like ER kinase (PERK); the inositol-requiring ER to nucleus signal kinase-1 (IRE1) and the activating transcription factor-6 (ATF6) (for review see e.g., Refs. [124,255]). It has been demonstrated that ricin inhibits activation of the UPR in yeast by preventing *Hac1* mRNA splicing [102]. The *Hac1* mRNA is an important regulator of the IRE1 signaling pathway. Activated yeast Ire1p triggers unconventional splicing of the *Hac1* mRNA leading to the synthesis of a transcription factor that specifically binds to promoters containing unfolded protein response elements [256]. Moreover, it was shown that RTA-mutated forms that could depurinate ribosomes but did not cause yeast cell death were unable to inhibit activation of the UPR by the ER stress-inducer tunicamycin [102]. These results suggest that the inability to activate the UPR in response to the ER stress contributes to the cytotoxicity of ricin. Other investigations also showed that ricin A-chain enhanced its own cytotoxicity by inhibiting the UPR [237]. In human epithelial cell lines (HeLa and MAC-T), RTA inhibited both phosphorylation of IRE1 and splicing of *XBP1* mRNA (homologue of yeast *Hac1*) induced by tunicamycin. However, in contrast to these studies, it was demonstrated that ricin can activate PERK and ATF6 of the UPR pathways, but not the IRE1 branch [238]. This led to cell growth arrest and apoptosis. It was proposed that blocking of the UPR response allows RTA to trigger cell death through a mechanism that is independent of protein synthesis inhibition.

3.3.4. Direct Action of Ricin on DNA and Ricin-Mediated Inhibition of DNA Repair Enzymes

As already mentioned in this review, not only rRNA but also DNA is a ricin substrate in the catalytic reaction mediated by this toxin [212]. It has been demonstrated that ricin and other RIPs can act on DNA and many different polynucleotidic substrates, releasing adenine from the sugar phosphate backbone of poly- and polydeoxynucleotides [257]. It was even suggested that RIPs should be classified as polynucleotide:adenosine glycosidases. The nuclear DNA injury revealed in cultured cells by the alkaline-halo assay and the alkaline filter elution technique was attributed to adenine release from RNA-free chromatin [212] and naked DNA [257]. Interestingly, in the case of ricin, the DNA damage was observed very early after cells treatment with ricin, and this damage was concomitant with the protein synthesis inhibition. At this time, the annexin V binding assay, caspase-3 activity, changes in cell morphology, and the formation of typical apoptotic DNA fragments were not detectable. It was suggested that ricin damages DNA in a way that does not result from ribosome inactivation or apoptosis [212] (Figure 4).

Ricin can also act indirectly on DNA by the inhibition of a DNA repair pathway. It has been demonstrated that this toxin, as well as other RIPs, releases an adenine from the ADP-ribosyl group of PARP [239]. NAD^+-dependent auto ADP-ribosylation of PARP is necessary for its active involvement in the DNA repair pathway called base excision repair [258]. It was suggested that depurination of auto-modified PARP by ricin results not only in the inhibition of DNA repair, but also leads to further ADP-ribosylation of PARP and depletion of the intracellular levels of NAD^+ and ATP. This, together with the impaired repair of damaged DNA, would cause cell necrosis induced by lethal amounts of ricin [239]. Moreover, it has been reported that ricin can inhibit the repair of H_2O_2 and the alkylating agent methyl methane sulphonate (MMS)-induced DNA lesions in HUVEC and U937 cells [259]. The inhibition of DNA repair by ricin seems to result from direct interactions with the DNA repair machinery. Importantly, ricin concentration used in the experiments to inhibit DNA repair was not sufficient to cause direct DNA damage or to induce total protein synthesis inhibition.

3.3.5. Ricin-Mediated Reactive Oxygen Species Production

The production of the reactive oxygen species (ROS) in cells has been reported to be involved in apoptosis induction by activating signal transduction mechanisms located upstream of the caspase-3 signalization pathway [260]. These pathways may be regulated by the changes in the oxidation status of the proteins involved in apoptosis signaling. It has been demonstrated that ricin increases the ROS levels in both yeast [24] and human cells [218]. The studies carried out on yeast suggested that the production of ROS is a necessary and sufficient condition for ricin-mediated induction of apoptosis [24]. Free radicals are scavenged by reduced glutathione (GSH). However, it was demonstrated in ricin-treated HeLa [218] and U937 cells [240] that the level of GSH was decreased. Thus, it was suggested that the GSH loss takes place downstream of caspase activation during the ricin-induced apoptotic process [240].

Interestingly, it has been demonstrated that the ROS formation is dependent on the presence of both extracellular and intracellular Ca^{2+} [261]. A rapid elevation of cellular calcium levels was observed in ricin-treated hepatoma cells [224]. Madin-Darby Canine Kidney (MDCK) cell death was significantly blocked by 1,9-deoxyforskolin (DDF) treatment, a drug that can reduce ion flux through several ion channels. This protective effect was significantly reversed by the increase in the extracellular Ca^{2+} concentrations [241].

Ricin-induced apoptosis is correlated not only with the elevation of the level of calcium ions. It was demonstrated that in ricin-treated U937 cells, the level of intracellular Zn^{2+} was increased and zinc was much more redistributed into the cytosol [242]. This occurs as an early apoptotic event, and exogenously-added Zn^{2+} inhibited the ricin-induced apoptosis. It was suggested that Zn^{2+} ions play a regulatory role in ricin-mediated apoptosis through their dissociation/association with certain intracellular elements.

3.3.6. Ricin B-Chain-Induced Apoptosis

Studies carried on U937 cells have demonstrated that the interaction of ricin B-chain with membrane glycoproteins and glycolipids may trigger signaling events leading to apoptosis [33]. This lectin activity-dependent mechanism was distinct from apoptosis signaling pathways induced by ricin A-chain. It has been demonstrated that carboxymethylated-(CM-) ricin B-chain was responsible for DNA fragmentation and typical apoptotic nuclear morphological changes, which were very similar to those observed in ricin-treated cells [33]. CM-ricin B-chain failed to inhibit protein synthesis in U937 cells. Thus, these experiments support the hypothesis that ricin-induced apoptosis or at least some of the apoptotic pathways are independent and not correlated with protein synthesis inhibition, at least in U937 cells. Recently published results have also shown that ricin-induced apoptosis is not solely attributed to the A-chain [243]. The intact heterodimeric ricin and ricin chains were injected into rats in order to study ricin-induced apoptosis in liver, which is a major site of in vivo ricin uptake and cytotoxicity [262]. It has been demonstrated that ricin was responsible for the intrinsic apoptosis pathway since increased cytochrome c content, activation of caspase-9 and caspase-3, and enrichment of DNA fragments in the cytosol were observed [243]. These authors observed also the B-chain in the cytosol and reported that it caused cytochrome c release from mitochondria in vivo and in vitro. These results suggest that a direct interaction of ricin B-chain with the mitochondrial outer membrane can be involved in ricin-induced apoptosis. The involvement of recombinant RTB in macrophage activation has also been studied [263]. It was demonstrated that RTB stimulated inducible nitric oxide (NO) synthase (iNOS) and TNF-α and IL-6 expression, which are involved in the activation of protein tyrosine kinase, NF-κB and JAK-STAT signaling.

4. Perspectives

4.1. Ricin-Based Immunotoxins

In the 19th century, Paul Ehrlich proposed the "magic bullet concept", which states that drugs can directly enter target cells and hit only abnormal cells of the human body [264]. Since then, the idea of a selective action of drugs that are able to affect only specific types of cells has been dynamically developed. However, the specificity of this process is challenging. The concept is utilized particularly in the field of cancer therapy with the application of immunotoxins (ITs) [265–268]. ITs are chimeric proteins composed of a toxin or a part of the toxin conjugated with a monoclonal antibody (mAb) or its fragment. When toxins are coupled to other carriers: growth factors, hormones or lectins that preferentially bind to some cell types, they are more commonly referred to as "chimeric toxins" or "conjugates" [269]. Ricin is the most commonly used plant toxin in the construction of ITs [30,269]. The first ricin-based ITs were prepared by binding holotoxin to a specific mAb [270]. Despite high efficiency, a large non-specific toxic effect of these immunotoxins was observed and made them impossible to use in a clinical setting. In a different experimental approach, only the A-chain of ricin was used, but also such ITs exhibited non-specific toxicity [271], due to the fact that receptors present on many cell types can recognize mannose residues present on the RTA [272]. To solve this problem, new ITs were prepared with deglycosylated RTA (dgRTA) [269,273,274].

Ricin-based immunotoxins are promising in the treatment of many types of diseases. Autoimmunity, immunodeficiency and neoplasia are examples of diseases connected with deregulation of the immune system. These dysfunctions are characterized by changes in the normal amount or function of Th (helper) cells. ITs composed of RTA and cell-reactive antibodies can specifically target neoplastic cells. It was shown that treatment of Th cells with Fab' fragments of anti-L3T4 antibody bound with RTA (Fab'anti-L3T4-A) inhibit keyhole limpet hemocyanin (KLH)-specific Th cells from proliferation and differentiation of the antigen-specific B cells (trinitrophenyl-(TNP)-specific B cells) [275]. These results indicate that Fab'anti-L3T4-A is able to specially inhibit Th cells that activate B cells. Another immunotoxin RTA-4D5-KDEL was constructed by connecting the anti-HER2 single chain variable fragment 4D5 scFv and KDEL, the ER-targeting peptides, with the C-terminal part of the RTA. Experiments showed that RTA-4D5-KDEL had a strong inhibitory effect on the ovarian cancer cells, SKOV-3, which were HER-2 overexpressing, and caused little damage to H460 lung cancer cells and to kidney HEK 293 cells. The KDEL of the RTA-4D5-KDEL immunotoxin was able to direct the recombinant protein to the ER. In light of this information, it can be assumed that this immunotoxin has a strong inhibitory effect on ovarian cancer cells with overexpression of HER2, and that it will exhibit little toxicity in normal cells [44]. Bladder cancer is one of the most frequent tumors. This disease is treated with transurethral resections and additionally with local immunotherapy or chemotherapy with good results; however, there is no ideal therapy to heal invasive carcinoma. A new antibody-based immunotoxin BCMab1-Ra was generated by linking of BCMab1-, a novel mouse monoclonal antibody, specific for aberrantly glycosylated Integrin a3b1 in this type of cancer with the ricin A-chain (Ra) [276]. The effect of the BCMab1-Ra on bladder cancer was investigated on a 57-year-old patient that refused radical surgery and chemotherapy. It has been demonstrated that the use of BCMab1-Ra first reduced the tumor, and that after 30 weeks of treatment there was no tumor observed by cystoscope examination. Moreover, human anti-mouse antibody (HAMA) that would indicate a strong immunologic response was not detectable in the blood circulation of this patient [276].

Various therapies are being utilized in the treatment of cancer. Traditional procedures such as radiation therapy, chemotherapy and surgery have some limitations and give serious side effects. Immunotoxins represent another technique with the possibility to increase the selectivity of action, but further development in this field is required.

4.2. Ricin Conjugated with Nanoparticles

During recent years, nanoparticles (NPs) have been studied intensively both as carriers used for delivery of therapeutic drugs (conventional drugs, recombinant proteins, vaccines and nucleotides) to certain cells and as therapeutic agents that may act per se or modulate activity of other compounds [277–280]. One carrier that can deliver (NPs) to cells is the ricin B-chain. The internalization mechanism as well as intracellular transport of ricinB:Quantum dot (QD) nanoparticle conjugates have been studied in different cells [41,281,282]. It was concluded that Qdots may have severe consequences on cell physiology [281,282]. Moreover, the internalization of ricinB:QDs in HeLa cells is dependent on dynamin and based on a macropinocytosis-like mechanism [41].

A complex of carbon dots (CDs) with RTB has been evaluated for enhanced immunomodulatory activity of RTB [283]. It was demonstrated that CDs-RTB can facilitate macrophage proliferation and increase the generation of nitric oxide (NO), IL-6 and TNF-α in RAW 264.7 cells, indicating enhanced immunomodulatory activity of CDs-RTB in comparison to RTB acting alone [283].

Another interesting example is a recombinant version of a ricin nanoparticle (T22-mRTA-H6) containing the T22 peptide, an efficient ligand of the cell surface marker CXCR4 (a cytokine receptor selectively overexpressed in metastatic cells of many cancer types) at the amino terminus followed by a mutated version of the ricin A chain and a hexahistidine tail at the carboxy terminus [284]. In this construct, mutation N132A was introduced to suppress the vascular leak syndrome (see below); a furin cleavage site was incorporated to allow the release of the N-terminal region in the endosome as well as a KDEL motif was added to mediate retrograde transport. Interestingly, this construct was engineered in order to allow for ricin A-chain aggregation and to become a targeting agent for the precise tumor delivery of protein-only nanoparticles. The recombinant T22-mRTA-H6 was produced in *E. coli* and purified. The spontaneous formation of self-assembled nanoparticles was possibly due to the combination of the cationic peptides at the amino terminus and polyhistidines at the carboxy terminus. T22-mRTA-H6 nanoparticles show highly selective therapeutic effects, and ricin A-chain was highly active on target cells, significantly reducing the effect of leukemia cells on relevant organs.

4.3. Vaccines against Ricin and Neutralizing Antibodies against Ricin

Despite numerous medical applications in which ricin can be used, this toxin is among the most potent and lethal substances that are known [35,285]. Currently, no approved vaccine or therapeutics exist to protect against ricin intoxication. The idea to develop a preventive vaccine against ricin has grown over the last years mainly because of the increasing concern that crude ricin powder can easily be made and used as a bio-threat agent. Two of the leading vaccine antigen candidates, the closely related RTA-based subunit vaccines, RiVax™ and RVEc™, are now under development [39,286,287]. RiVax™ is a full-length recombinant derivative of RTA whose enzymatic activity has been largely eliminated through a point mutation in a key active site residue (Y80A). RiVax™ also contains a mutation in the site (V76M) attributed to the induction of the vascular leak syndrome (VLS) [286]. The VLS is the main side-effect of ricin-derived immunotoxins. It has a complex etiology involving damage to vascular endothelial cells [288]. Mutation in the vaccine to alter the VLS motif was introduced to eliminate this toxicity. RVEc™ is a truncated derivative of RTA that lacks the hydrophobic carboxy-terminal region (residues 199–267) as well as a small hydrophobic loop in the N-terminus (residues 34–43). RVEc™ mutations do not directly influence the active site of RTA, but the removal of both regions causes that ricin present in this vaccine is inactive with reduced ability to cause the vascular leak syndrome [39,287]. Both candidate vaccines are under investigation in animal studies and Phase I clinical trials. Furthermore, the results of two Phase I clinical trials have indicated that RiVax™ is safe and immunogenic in humans [289,290]. One obvious strategy to augment the overall immunogenicity of vaccines is the use of next-generation adjuvants. However, adjuvants themselves may not be sufficient to achieve maximal immunogenicity. Enhancing the immunogenicity of vaccines may require a structure-based redesign of the antigen itself. The resulting combinations of mutations led to the identification of derivatives of RiVax which are several times more efficient [291].

In the late 1880s, Paul Erhlich and others first described the potential use of antibodies (Abs) to completely inactivate the toxin. Immunity to ricin is associated with the production of protective antibodies. Since those early studies, many studies of antisera and antibody preparations derived from different animal species and tested on a diversity of cell types have been made. Anti-RTA and RTB antibodies were tested in rabbits and mice and displayed some neutralization action. Some results suggest that antibodies neutralize ricin by perturbing toxin uptake and/or intracellular trafficking without affecting the toxin attachment to cell surfaces. This confers passive immunity in vivo [37,291–293]. On the other hand, blocking ricin attachment to receptors on the cell surface is the mechanisms of action of other specific antibodies (24B11 and VHH D10/B7) [294]. Multiple studies revealed that there are three general classes of ricin-specific Abs; those that bind RTA, RTB and ricin holotoxin [286,295]. As part of an effort to engineer ricin antitoxin and immunotherapies, libraries of phage-displayed, heavy chain-only antibodies (V$_H$Hs) have been produced and well characterized [296]. It has been demonstrated that immunity against ricin is mediated by antibody. However, the specificity of particular epitopes involved in protective immunity remains unclear. The importance of toxin-neutralizing antibodies in protection against ricin is not questioned. However, the exact correlation between the structure of RTA and the induction of protective immunity must be more strictly evaluated. In just the past 10 years, several reports have been published that demonstrated that passive administration of a toxin-neutralizing antibody is sufficient to display mice protection to a lethal dose of ricin delivered by injection, ingestion or inhalation [292,297–301].

5. Concluding Remarks

Ricin can be considered as a powerful tool to study intracellular pathways in general and cell death mechanisms that include apoptosis, inflammation or cell stress-induced signaling. On the other hand, exact knowledge about ricin action in the target cells is necessary to produce effectively working ricin-based immunotoxins or vaccines. One of the most interesting discoveries that has been made recently describes a vital sugar code for ricin toxicity, that is conserved from mouse to human [100]. This mechanism is based on defined glycosidic structures that determine cellular fate upon exposure to the toxin. The question of whether depurination of rRNA is necessary for ricin-induced cell death is still being discussed. It is believed that in addition to rRNA damage, ricin can induce apoptosis, inflammation and DNA damage. The correlation between these processes has been intensively studied. This knowledge is constantly being expanded as the huge contribution to this field has been made over the past years. An old dogma about ricin has currently been investigated. It was believed that a single A-chain molecule of ricin or other type-2 RIPs have the ability to kill one eukaryotic cell [20]. However, it was recently reported that one or a few molecules of ricin A-chain present in the cytosol is not sufficient to inhibit protein synthesis [302]. Moreover, cells with a partial inhibition of protein synthesis can, upon ricin removal, increase the level of protein production and survive the toxin challenge. Thus, in contrast to the previously accepted model, ongoing toxin delivery to the cytosol appears to be necessary for the death of cells exposed to sub-optimal ricin concentrations [302]. It was also suggested that ricin and other RIPs can be more toxic to cancer cells than to normal cells, due to the higher rate of protein synthesis in malignant cells during proliferation or due to the changes in receptor concentration on their surfaces or altered intracellular transport of the toxin [30,42,303–305]. This is, however, not always the case (for review see e.g., Refs. [30,285]). Thus, for specific delivery of ricin to cancer cells, directing ricin to particular epitopes on tumor cells is necessary. On the other hand, some specific properties of ricin may enhance its effect on cancer cells. The ability of RTA to inhibit UPR may make it more potent in targeted therapy for cancer. It has been demonstrated that an increased level of spliced *XBP1* relatively to unspliced *XBP1* correlates with poor prognosis in breast cancer [306], and XBP1 has been proposed as a therapeutic target for solid tumors [307]. Thus, ricin treatment may be particularly useful in cancer cells where UPR is already accelerated by conditions such as hypoxia. These findings highlight the role of ricin as a valuable component of modern immunotoxins.

Funding: The work referred to as from the Słomińska-Wojewódzka group was supported by the National Science Centre Poland grant 2015/19/B/NZ3/03266. Work with toxins in the group of Sandvig is supported by the Norwegian Cancer Society and the Southern and Eastern Norway Regional Health Authority.

Conflicts of Interest: The authors declare no conflict of interest.

References

1. Stillmark, H. Ueber Ricin, ein giftiges Ferment aus den Samen von *Ricinus comm.* L. und einigen anderen Euphorbiaceen: Inaugural-Dissertation. MD Thesis, University of Dorpat, Dorpat, Estonia, 1888.
2. Boyd, W.C.; Shapleigh, E. Diagnosis by subgroups of blood groups A and AB by use of plant agglutinins (lectins). *J. Lab. Clin. Med.* **1954**, *44*, 235–237. [PubMed]
3. Olsnes, S.; Pihl, A. Ricin—A potent inhibitor of protein synthesis. *FEBS Lett.* **1972**, *20*, 327–329. [CrossRef]
4. Olsnes, S.; Pihl, A. Different biological properties of the two constituent peptide chains of ricin, a toxic protein inhibiting protein synthesis. *Biochemistry* **1973**, *12*, 3121–3126. [CrossRef] [PubMed]
5. Olsnes, S.; Heiberg, R.; Pihl, A. Inactivation of eucaryotic ribosomes by the toxic plant proteins abrin and ricin. *Mol. Biol. Rep.* **1973**, *1*, 15–20. [CrossRef] [PubMed]
6. Endo, Y.; Tsurugi, K. RNA N-glycosidase activity of ricin A-chain. Mechanism of action of the toxic lectin ricin on eukaryotic ribosomes. *J. Biol. Chem.* **1987**, *262*, 8128–8130.
7. Endo, Y.; Mitsui, K.; Motizuki, M.; Tsurugi, K. The mechanism of action of ricin and related toxic lectins on eukaryotic ribosomes. The site and the characteristics of the modification in 28 S ribosomal RNA caused by the toxins. *J. Biol. Chem.* **1987**, *262*, 5908–5912.
8. Lord, J.M.; Roberts, L.M.; Robertus, J.D. Ricin: Structure, mode of action, and some current applications. *FASEB J.* **1994**, *8*, 201–208. [CrossRef]
9. Olsnes, S.; Refsnes, K.; Pihl, A. Mechanism of action of the toxic lectins abrin and ricin. *Nature* **1974**, *249*, 627–631. [CrossRef]
10. Shi, W.-W.; Mak, A.N.-S.; Wong, K.-B.; Shaw, P.-C. Structures and Ribosomal Interaction of Ribosome-Inactivating Proteins. *Molecules* **2016**, *21*, 1588. [CrossRef]
11. Zhou, Y.; Li, X.-P.; Kahn, J.N.; Tumer, N.E. Functional Assays for Measuring the Catalytic Activity of Ribosome Inactivating Proteins. *Toxins* **2018**, *10*, 240. [CrossRef]
12. Barbieri, L.; Battelli, M.G.; Stirpe, F. Ribosome-inactivating proteins from plants. *Biochim. Biophys. Acta* **1993**, *1154*, 237–282. [CrossRef]
13. Voorhees, R.M.; Schmeing, T.M.; Kelley, A.C.; Ramakrishnan, V. The mechanism for activation of GTP hydrolysis on the ribosome. *Science* **2010**, *330*, 835–838. [CrossRef] [PubMed]
14. Shi, X.; Khade, P.K.; Sanbonmatsu, K.Y.; Joseph, S. Functional role of the sarcin-ricin loop of the 23S rRNA in the elongation cycle of protein synthesis. *J. Mol. Biol.* **2012**, *419*, 125–138. [CrossRef] [PubMed]
15. Lancaster, L.; Lambert, N.J.; Maklan, E.J.; Horan, L.H.; Noller, H.F. The sarcin–ricin loop of 23S rRNA is essential for assembly of the functional core of the 50S ribosomal subunit. *RNA* **2008**, *14*, 1999–2012. [CrossRef] [PubMed]
16. Moazed, D.; Robertson, J.M.; Noller, H.F. Interaction of elongation factors EF-G and EF-Tu with a conserved loop in 23S RNA. *Nature* **1988**, *334*, 362–364. [CrossRef] [PubMed]
17. Hartley, M.R.; Lord, J.M. Cytotoxic ribosome-inactivating lectins from plants. *Biochim. Biophys. Acta* **2004**, *1701*, 1–14. [CrossRef] [PubMed]

18. Olsnes, S.; Fernandez-Puentes, C.; Carrasco, L.; Vazquez, D. Ribosome inactivation by the toxic lectins abrin and ricin. Kinetics of the enzymatic activity of the toxin A-chains. *Eur. J. Biochem.* **1975**, *60*, 281–288. [CrossRef] [PubMed]

19. Olsnes, S. The history of ricin, abrin and related toxins. *Toxicon* **2004**, *44*, 361–370. [CrossRef] [PubMed]

20. Eiklid, K.; Olsnes, S.; Pihl, A. Entry of lethal doses of abrin, ricin and modeccin into the cytosol of HeLa cells. *Exp. Cell Res.* **1980**, *126*, 321–326. [CrossRef]

21. Olson, K.R. *Poisoning and Drug Overdose, Sixth Edition*, 6th ed.; McGraw-Hill Education/Medical: New York, NY, USA, 2011; ISBN 978-0-07-166833-0.

22. Moshiri, M.; Hamid, F.; Etemad, L. Ricin Toxicity: Clinical and Molecular Aspects. *Rep. Biochem. Mol. Biol.* **2016**, *4*, 60–65.

23. *Convention on the Prohibition of the Development, Production, Stockpiling and Use of Chemical Weapons and on Their Destruction*; Organisation for the Probhibition of Chemical Weapons (OPCW): The Hague, The Netherlands, 2005.

24. Li, X.-P.; Baricevic, M.; Saidasan, H.; Tumer, N.E. Ribosome depurination is not sufficient for ricin-mediated cell death in Saccharomyces cerevisiae. *Infect. Immun.* **2007**, *75*, 417–428. [CrossRef] [PubMed]

25. Flexner, S. The histological changes produced by ricin and abrin intoxications. *J. Exp. Med.* **1897**, *2*, 197–216. [CrossRef] [PubMed]

26. Griffiths, G.D.; Leek, M.D.; Gee, D.J. The toxic plant proteins ricin and abrin induce apoptotic changes in mammalian lymphoid tissues and intestine. *J. Pathol.* **1987**, *151*, 221–229. [CrossRef] [PubMed]

27. Hughes, J.N.; Lindsay, C.D.; Griffiths, G.D. Morphology of ricin and abrin exposed endothelial cells is consistent with apoptotic cell death. *Hum. Exp. Toxicol.* **1996**, *15*, 443–451. [CrossRef] [PubMed]

28. Day, P.J.; Pinheiro, T.J.T.; Roberts, L.M.; Lord, J.M. Binding of ricin A-chain to negatively charged phospholipid vesicles leads to protein structural changes and destabilizes the lipid bilayer. *Biochemistry* **2002**, *41*, 2836–2843. [CrossRef] [PubMed]

29. Kumar, O.; Sugendran, K.; Vijayaraghavan, R. Oxidative stress associated hepatic and renal toxicity induced by ricin in mice. *Toxicon* **2003**, *41*, 333–338. [CrossRef]

30. Słomińska-Wojewódzka, M.; Sandvig, K. Ricin and Ricin-Containing Immunotoxins: Insights into Intracellular Transport and Mechanism of action in Vitro. *Antibodies* **2013**, *2*, 236–269. [CrossRef]

31. Shih, S.-F.; Wu, Y.-H.; Hung, C.-H.; Yang, H.-Y.; Lin, J.-Y. Abrin Triggers Cell Death by Inactivating a Thiol-specific Antioxidant Protein. *J. Biol. Chem.* **2001**, *276*, 21870–21877. [CrossRef] [PubMed]

32. Baluna, R.; Coleman, E.; Jones, C.; Ghetie, V.; Vitetta, E.S. The Effect of a Monoclonal Antibody Coupled to Ricin A Chain-Derived Peptides on Endothelial Cells in Vitro: Insights into Toxin-Mediated Vascular Damage. *Exp. Cell Res.* **2000**, *258*, 417–424. [CrossRef]

33. Hasegawa, N.; Kimura, Y.; Oda, T.; Komatsu, N.; Muramatsu, T.H.-J. Isolated ricin B-chain-mediated apoptosis in U937 cells. *Biosci. Biotechnol. Biochem.* **2000**, *64*, 1422–1429. [CrossRef]

34. Berger, T.; Eisenkraft, A.; Bar-Haim, E.; Kassirer, M.; Aran, A.A.; Fogel, I. Toxins as biological weapons for terror-characteristics, challenges and medical countermeasures: A mini-review. *Disaster Mil. Med.* **2016**, *2*, 7. [CrossRef] [PubMed]

35. Janik, E.; Ceremuga, M.; Saluk-Bijak, J.; Bijak, M. Biological Toxins as the Potential Tools for Bioterrorism. *Int. J. Mol. Sci.* **2019**, *20*, 1181. [CrossRef] [PubMed]

36. Hu, W.; Yin, J.; Chau, D.; Hu, C.C.; Cherwonogrodzky, J.W. Anti-Ricin Protective Monoclonal Antibodies. *Ricin Toxin* **2014**, *14*, 145–158.

37. Herrera, C.; Klokk, T.I.; Cole, R.; Sandvig, K.; Mantis, N.J. A Bispecific Antibody Promotes Aggregation of Ricin Toxin on Cell Surfaces and Alters Dynamics of Toxin Internalization and Trafficking. *PLoS ONE* **2016**, *11*, e0156893. [CrossRef]

38. Yermakova, A.; Klokk, T.I.; O'Hara, J.M.; Cole, R.; Sandvig, K.; Mantis, N.J. Neutralizing Monoclonal Antibodies against Disparate Epitopes on Ricin Toxin's Enzymatic Subunit Interfere with Intracellular Toxin Transport. *Sci. Rep.* **2016**, *6*, 22721. [CrossRef] [PubMed]

39. Brey, R.N.; Mantis, N.J.; Pincus, S.H.; Vitetta, E.S.; Smith, L.A.; Roy, C.J. Recent advances in the development of vaccines against ricin. *Hum. Vaccin Immunother.* **2016**, *12*, 1196–1201. [CrossRef] [PubMed]

40. Liu, X.; Pop, L.M.; Schindler, J.; Vitetta, E.S. Immunotoxins constructed with chimeric, short-lived anti-CD22 monoclonal antibodies induce less vascular leak without loss of cytotoxicity. *MAbs* **2012**, *4*, 57–68. [CrossRef] [PubMed]

41. Iversen, T.G.; Frerker, N.; Sandvig, K. Uptake of ricinB-quantum dot nanoparticles by a macropinocytosis-like mechanism. *J. Nanobiotechnol.* **2012**, *10*, 33. [CrossRef] [PubMed]

42. Tyagi, N.; Tyagi, M.; Pachauri, M.; Ghosh, P.C. Potential therapeutic applications of plant toxin-ricin in cancer: Challenges and advances. *Tumor Biol.* **2015**, *36*, 8239–8246. [CrossRef]

43. Pizzo, E.; Di Maro, A. A new age for biomedical applications of Ribosome Inactivating Proteins (RIPs): From bioconjugate to nanoconstructs. *J. Biomed Sci.* **2016**, *23*, 54. [CrossRef]

44. Jiao, P.; Zhang, J.; Dong, Y.; Wei, D.; Ren, Y. Construction and characterization of the recombinant immunotoxin RTA-4D5-KDEL targeting HER2/neu-positive cancer cells and locating the endoplasmic reticulum. *Appl. Microbiol. Biotechnol.* **2018**, *102*, 9585–9594. [CrossRef]

45. Magnusson, S.; Kjeken, R.; Berg, T. Characterization of two distinct pathways of endocytosis of ricin by rat liver endothelial cells. *Exp. Cell Res.* **1993**, *205*, 118–125. [CrossRef]

46. Sphyris, N.; Lord, J.M.; Wales, R.; Roberts, L.M. Mutational analysis of the Ricinus lectin B-chains. Galactose-binding ability of the 2 gamma subdomain of Ricinus communis agglutinin B-chain. *J. Biol. Chem.* **1995**, *270*, 20292–20297. [CrossRef] [PubMed]

47. Sandvig, K.; Olsnes, S.; Pihl, A. Kinetics of binding of the toxic lectins abrin and ricin to surface receptors of human cells. *J. Biol. Chem.* **1976**, *251*, 3977–3984.

48. Sandvig, K.; Spilsberg, B.; Lauvrak, S.U.; Torgersen, M.L.; Iversen, T.-G.; van Deurs, B.O. Pathways followed by protein toxins into cells. *Int. J. Med. Microbiol.* **2004**, *293*, 483–490. [CrossRef] [PubMed]

49. Van Deurs, B.; Pedersen, L.R.; Sundan, A.; Olsnes, S.; Sandvig, K. Receptor-mediated endocytosis of a ricin-colloidal gold conjugate in vero cells. Intracellular routing to vacuolar and tubulo-vesicular portions of the endosomal system. *Exp. Cell Res.* **1985**, *159*, 287–304. [CrossRef]

50. Moya, M.; Dautry-Varsat, A.; Goud, B.; Louvard, D.; Boquet, P. Inhibition of coated pit formation in Hep2 cells blocks the cytotoxicity of diphtheria toxin but not that of ricin toxin. *J. Cell Biol.* **1985**, *101*, 548–559. [CrossRef] [PubMed]

51. Sandvig, K.; Olsnes, S.; Petersen, O.W.; van Deurs, B. Inhibition of endocytosis from coated pits by acidification of the cytosol. *J. Cell. Biochem.* **1988**, *36*, 73–81. [CrossRef] [PubMed]

52. Simpson, J.C.; Smith, D.C.; Roberts, L.M.; Lord, J.M. Expression of mutant dynamin protects cells against diphtheria toxin but not against ricin. *Exp. Cell Res.* **1998**, *239*, 293–300. [CrossRef]

53. Spooner, R.A.; Lord, J.M. Ricin Trafficking in Cells. *Toxins* **2015**, *7*, 49–65. [CrossRef] [PubMed]

54. Rodal, S.K.; Skretting, G.; Garred, O.; Vilhardt, F.; van Deurs, B.; Sandvig, K. Extraction of cholesterol with methyl-beta-cyclodextrin perturbs formation of clathrin-coated endocytic vesicles. *Mol. Biol. Cell* **1999**, *10*, 961–974. [CrossRef] [PubMed]

55. Sandvig, K.; Pust, S.; Skotland, T.; van Deurs, B. Clathrin-independent endocytosis: Mechanisms and function. *Curr. Opin. Cell Biol.* **2011**, *23*, 413–420. [CrossRef] [PubMed]

56. Sandvig, K.; Kavaliauskiene, S.; Skotland, T. Clathrin-independent endocytosis: An increasing degree of complexity. *Histochem. Cell Biol.* **2018**, *150*, 107–118. [CrossRef] [PubMed]

57. Sandvig, K.; Grimmer, S.; Iversen, T.G.; Rodal, K.; Torgersen, M.L.; Nicoziani, P.; van Deurs, B. Ricin transport into cells: Studies of endocytosis and intracellular transport. *Int. J. Med. Microbiol.* **2000**, *290*, 415–420. [CrossRef]

58. Sokołowska, I.; Wälchli, S.; Węgrzyn, G.; Sandvig, K.; Słomińska-Wojewódzka, M. A single point mutation in ricin A-chain increases toxin degradation and inhibits EDEM1-dependent ER retrotranslocation. *Biochem. J.* **2011**, *436*, 371–385. [CrossRef]

59. Sokołowska, I.; Piłka, E.S.; Sandvig, K.; Węgrzyn, G.; Słomińska-Wojewódzka, M. Hydrophobicity of protein determinants influences the recognition of substrates by EDEM1 and EDEM2 in human cells. *BMC Cell Biol.* **2015**, *16*, 1. [CrossRef]

60. Becker, B.; Schnöder, T.; Schmitt, M.J. Yeast Reporter Assay to Identify Cellular Components of Ricin Toxin A Chain Trafficking. *Toxins* **2016**, *8*, 366. [CrossRef]

61. Lewis, M.J.; Pelham, H.R.B. A new yeast endosomal SNARE related to mammalian syntaxin 8. *Traffic* **2002**, *3*, 922–929. [CrossRef]

62. Tang, H.; Song, M.; He, Y.; Wang, J.; Wang, S.; Shen, Y.; Hou, J.; Bao, X. Engineering vesicle trafficking improves the extracellular activity and surface display efficiency of cellulases in Saccharomyces cerevisiae. *Biotechnol. Biofuels* **2017**, *10*, 53. [CrossRef]

63. Sandvig, K.; Olsnes, S.; Pihl, A. Interactions between abrus lectins and Sephadex particles possessing immobilized desialylated fetuin. Model studies of the interaction of lectins with cell surface receptors. *Eur. J. Biochem.* **1978**, *88*, 307–313. [CrossRef]

64. Blum, J.S.; Fiani, M.L.; Stahl, P.D. Proteolytic cleavage of ricin A chain in endosomal vesicles. Evidence for the action of endosomal proteases at both neutral and acidic pH. *J. Biol. Chem.* **1991**, *266*, 22091–22095. [PubMed]

65. Sandvig, K.; van Deurs, B. Membrane traffic exploited by protein toxins. *Annu. Rev. Cell Dev. Biol.* **2002**, *18*, 1–24. [CrossRef] [PubMed]

66. Van Deurs, B.; Tønnessen, T.I.; Petersen, O.W.; Sandvig, K.; Olsnes, S. Routing of internalized ricin and ricin conjugates to the Golgi complex. *J. Cell Biol.* **1986**, *102*, 37–47. [CrossRef] [PubMed]

67. Van Deurs, B.; Sandvig, K.; Petersen, O.W.; Olsnes, S.; Simons, K.; Griffiths, G. Estimation of the amount of internalized ricin that reaches the trans-Golgi network. *J. Cell Biol.* **1988**, *106*, 253–267. [CrossRef] [PubMed]

68. Sandvig, K.; van Deurs, B. Delivery into cells: Lessons learned from plant and bacterial toxins. *Gene Ther.* **2005**, *12*, 865–872. [CrossRef] [PubMed]

69. Rapak, A.; Falnes, P.O.; Olsnes, S. Retrograde transport of mutant ricin to the endoplasmic reticulum with subsequent translocation to cytosol. *Proc. Natl. Acad. Sci. USA* **1997**, *94*, 3783–3788. [CrossRef]

70. Driouich, A.; Zhang, G.F.; Staehelin, L.A. Effect of brefeldin A on the structure of the Golgi apparatus and on the synthesis and secretion of proteins and polysaccharides in sycamore maple (*Acer pseudoplatanus*) suspension-cultured cells. *Plant Physiol.* **1993**, *101*, 1363–1373. [CrossRef]

71. Yoshida, T.; Chen, C.C.; Zhang, M.S.; Wu, H.C. Disruption of the Golgi apparatus by brefeldin A inhibits the cytotoxicity of ricin, modeccin, and Pseudomonas toxin. *Exp. Cell Res.* **1991**, *192*, 389–395. [CrossRef]

72. Sandvig, K.; Prydz, K.; Hansen, S.H.; van Deurs, B. Ricin transport in brefeldin A-treated cells: Correlation between Golgi structure and toxic effect. *J. Cell Biol.* **1991**, *115*, 971–981. [CrossRef]

73. Prydz, K.; Hansen, S.H.; Sandvig, K.; van Deurs, B. Effects of brefeldin A on endocytosis, transcytosis and transport to the Golgi complex in polarized MDCK cells. *J. Cell Biol.* **1992**, *119*, 259–272. [CrossRef]

74. Leitinger, B.; Brown, J.L.; Spiess, M. Tagging secretory and membrane proteins with a tyrosine sulfation site. Tyrosine sulfation precedes galactosylation and sialylation in COS-7 cells. *J. Biol. Chem.* **1994**, *269*, 8115–8121.

75. Llorente, A.; Rapak, A.; Schmid, S.L.; van Deurs, B.; Sandvig, K. Expression of Mutant Dynamin Inhibits Toxicity and Transport of Endocytosed Ricin to the Golgi Apparatus. *J. Cell Biol.* **1998**, *140*, 553–563. [CrossRef] [PubMed]

76. Gray, J.L.; von Delft, F.; Brennan, P. Targeting the Small GTPase Superfamily through their Regulatory Proteins. *Angew. Chem. Int. Ed.* **2019**. [CrossRef]

77. Lombardi, D.; Soldati, T.; Riederer, M.A.; Goda, Y.; Zerial, M.; Pfeffer, S.R. Rab9 functions in transport between late endosomes and the trans Golgi network. *EMBO J.* **1993**, *12*, 677–682. [CrossRef] [PubMed]

78. Riederer, M.A.; Soldati, T.; Shapiro, A.D.; Lin, J.; Pfeffer, S.R. Lysosome biogenesis requires Rab9 function and receptor recycling from endosomes to the trans-Golgi network. *J. Cell Biol* **1994**, *125*, 573–582. [CrossRef] [PubMed]

79. Iversen, T.G.; Skretting, G.; Llorente, A.; Nicoziani, P.; van Deurs, B.; Sandvig, K. Endosome to Golgi transport of ricin is independent of clathrin and of the Rab9- and Rab11-GTPases. *Mol. Biol. Cell* **2001**, *12*, 2099–2107. [CrossRef]

80. Mallard, F.; Tang, B.L.; Galli, T.; Tenza, D.; Saint-Pol, A.; Yue, X.; Antony, C.; Hong, W.; Goud, B.; Johannes, L. Early/recycling endosomes-to-TGN transport involves two SNARE complexes and a Rab6 isoform. *J. Cell Biol.* **2002**, *156*, 653–664. [CrossRef] [PubMed]

81. Echard, A.; Opdam, F.J.; de Leeuw, H.J.; Jollivet, F.; Savelkoul, P.; Hendriks, W.; Voorberg, J.; Goud, B.; Fransen, J.A. Alternative splicing of the human Rab6A gene generates two close but functionally different isoforms. *Mol. Biol. Cell* **2000**, *11*, 3819–3833. [CrossRef]

82. Mallard, F.; Antony, C.; Tenza, D.; Salamero, J.; Goud, B.; Johannes, L. Direct pathway from early/recycling endosomes to the Golgi apparatus revealed through the study of shiga toxin B-fragment transport. *J. Cell Biol.* **1998**, *143*, 973–990. [CrossRef]

83. Utskarpen, A.; Slagsvold, H.H.; Iversen, T.-G.; Wälchli, S.; Sandvig, K. Transport of ricin from endosomes to the Golgi apparatus is regulated by Rab6A and Rab6A'. *Traffic* **2006**, *7*, 663–672. [CrossRef]

84. Grimmer, S.; Iversen, T.-G.; van Deurs, B.; Sandvig, K. Endosome to Golgi Transport of Ricin Is Regulated by Cholesterol. *Mol. Biol. Cell* **2000**, *11*, 4205–4216. [CrossRef] [PubMed]

85. Lauvrak, S.U.; Llorente, A.; Iversen, T.-G.; Sandvig, K. Selective regulation of the Rab9-independent transport of ricin to the Golgi apparatus by calcium. *J. Cell Sci.* **2002**, *115*, 3449–3456. [PubMed]

86. Porat, A.; Elazar, Z. Regulation of intra-Golgi membrane transport by calcium. *J. Biol. Chem.* **2000**, *275*, 29233–29237. [CrossRef] [PubMed]

87. Moreau, D.; Kumar, P.; Wang, S.C.; Chaumet, A.; Chew, S.Y.; Chevalley, H.; Bard, F. Genome-wide RNAi screens identify genes required for Ricin and PE intoxications. *Dev. Cell* **2011**, *21*, 231–244. [CrossRef] [PubMed]

88. Becker, B.; Schmitt, M.J. A Simple Fluorescence-based Reporter Assay to Identify Cellular Components Required for Ricin Toxin A Chain (RTA) Trafficking in Yeast. *J. Vis. Exp.* **2017**, *130*, e56588. [CrossRef] [PubMed]

89. Birkeli, K.A.; Llorente, A.; Torgersen, M.L.; Keryer, G.; Taskén, K.; Sandvig, K. Endosome-to-Golgi transport is regulated by protein kinase A type II alpha. *J. Biol. Chem.* **2003**, *278*, 1991–1997. [CrossRef] [PubMed]

90. Skånland, S.S.; Wälchli, S.; Utskarpen, A.; Wandinger-Ness, A.; Sandvig, K. Phosphoinositide-regulated retrograde transport of ricin: Crosstalk between hVps34 and sorting nexins. *Traffic* **2007**, *8*, 297–309. [CrossRef]

91. Sandvig, K.; van Deurs, B. Entry of ricin and Shiga toxin into cells: Molecular mechanisms and medical perspectives. *EMBO J.* **2000**, *19*, 5943–5950. [CrossRef]

92. Day, P.J.; Owens, S.R.; Wesche, J.; Olsnes, S.; Roberts, L.M.; Lord, J.M. An interaction between ricin and calreticulin that may have implications for toxin trafficking. *J. Biol. Chem.* **2001**, *276*, 7202–7208. [CrossRef]

93. Spooner, R.A.; Smith, D.C.; Easton, A.J.; Roberts, L.M.; Lord, J.M. Retrograde transport pathways utilised by viruses and protein toxins. *Virol. J.* **2006**, *3*, 26. [CrossRef]

94. Chen, A.; AbuJarour, R.J.; Draper, R.K. Evidence that the transport of ricin to the cytoplasm is independent of both Rab6A and COPI. *J. Cell Sci.* **2003**, *116*, 3503–3510. [CrossRef] [PubMed]

95. Llorente, A.; Lauvrak, S.U.; van Deurs, B.; Sandvig, K. Induction of direct endosome to endoplasmic reticulum transport in Chinese hamster ovary (CHO) cells (LdlF) with a temperature-sensitive defect in epsilon-coatomer protein (epsilon-COP). *J. Biol. Chem.* **2003**, *278*, 35850–35855. [CrossRef] [PubMed]

96. Otte, S.; Belden, W.J.; Heidtman, M.; Liu, J.; Jensen, O.N.; Barlowe, C. Erv41p and Erv46p: New components of COPII vesicles involved in transport between the ER and Golgi complex. *J. Cell Biol.* **2001**, *152*, 503–518. [CrossRef] [PubMed]

97. Adolf, F.; Rhiel, M.; Hessling, B.; Gao, Q.; Hellwig, A.; Béthune, J.; Wieland, F.T. Proteomic Profiling of Mammalian COPII and COPI Vesicles. *Cell Rep.* **2019**, *26*, 250–265. [CrossRef]

98. Sato, K.; Sato, M.; Nakano, A. Rer1p, a Retrieval Receptor for Endoplasmic Reticulum Membrane Proteins, Is Dynamically Localized to the Golgi Apparatus by Coatomer. *J. Cell Biol.* **2001**, *152*, 935–944. [CrossRef] [PubMed]

99. Liu, Y.; Flanagan, J.J.; Barlowe, C. Sec22p export from the endoplasmic reticulum is independent of SNARE pairing. *J. Biol. Chem.* **2004**, *279*, 27225–27232. [CrossRef]

100. Taubenschmid, J.; Stadlmann, J.; Jost, M.; Klokk, T.I.; Rillahan, C.D.; Leibbrandt, A.; Mechtler, K.; Paulson, J.C.; Jude, J.; Zuber, J.; et al. A vital sugar code for ricin toxicity. *Cell Res.* **2017**, *27*, 1351–1364. [CrossRef] [PubMed]

101. Simpson, J.C.; Roberts, L.M.; Römisch, K.; Davey, J.; Wolf, D.H.; Lord, J.M. Ricin A chain utilises the endoplasmic reticulum-associated protein degradation pathway to enter the cytosol of yeast. *FEBS Lett.* **1999**, *459*, 80–84. [CrossRef]

102. Parikh, B.A.; Tortora, A.; Li, X.-P.; Tumer, N.E. Ricin inhibits activation of the unfolded protein response by preventing splicing of the HAC1 mRNA. *J. Biol. Chem.* **2008**, *283*, 6145–6153. [CrossRef]

103. Spooner, R.A.; Watson, P.D.; Marsden, C.J.; Smith, D.C.; Moore, K.A.H.; Cook, J.P.; Lord, J.M.; Roberts, L.M. Protein disulphide-isomerase reduces ricin to its A and B chains in the endoplasmic reticulum. *Biochem. J.* **2004**, *383*, 285–293. [CrossRef]

104. Redmann, V.; Oresic, K.; Tortorella, L.L.; Cook, J.P.; Lord, M.; Tortorella, D. Dislocation of Ricin Toxin A Chains in Human Cells Utilizes Selective Cellular Factors. *J. Biol. Chem.* **2011**, *286*, 21231–21238. [CrossRef]

105. Di Cola, A.; Frigerio, L.; Lord, J.M.; Ceriotti, A.; Roberts, L.M. Ricin A chain without its partner B chain is degraded after retrotranslocation from the endoplasmic reticulum to the cytosol in plant cells. *Proc. Natl. Acad. Sci. USA* **2001**, *98*, 14726–14731. [CrossRef] [PubMed]

106. Gemmill, T.R.; Trimble, R.B. Overview of N- and O-linked oligosaccharide structures found in various yeast species. *Biochim. Biophys. Acta Gen. Subj.* **1999**, *1426*, 227–237. [CrossRef]

107. Lamb, F.I.; Roberts, L.M.; Lord, J.M. Nucleotide sequence of cloned cDNA coding for preproricin. *Eur. J. Biochem.* **1985**, *148*, 265–270. [CrossRef] [PubMed]

108. Wesche, J.; Rapak, A.; Olsnes, S. Dependence of ricin toxicity on translocation of the toxin A-chain from the endoplasmic reticulum to the cytosol. *J. Biol. Chem.* **1999**, *274*, 34443–34449. [CrossRef]

109. Spooner, R.A.; Lord, J.M. How ricin and Shiga toxin reach the cytosol of target cells: Retrotranslocation from the endoplasmic reticulum. *Curr. Top. Microbiol. Immunol.* **2012**, *357*, 19–40.

110. Nowakowska-Gołacka, J.; Sominka, H.; Sowa-Rogozińska, N.; Słomińska-Wojewódzka, M. Toxins Utilize the Endoplasmic Reticulum-Associated Protein Degradation Pathway in Their Intoxication Process. *Int. J. Mol. Sci.* **2019**, *20*, 1307. [CrossRef]

111. Mohanraj, D.; Ramakrishnan, S. Cytotoxic effects of ricin without an interchain disulfide bond: Genetic modification and chemical crosslinking studies. *Biochim. Biophys. Acta Gen. Subj.* **1995**, *1243*, 399–406. [CrossRef]

112. Freedman, R.B.; Hirst, T.R.; Tuite, M.F. Protein disulphide isomerase: Building bridges in protein folding. *Trends Biochem. Sci.* **1994**, *19*, 331–336. [CrossRef]

113. Bulleid, N.J. Disulfide Bond Formation in the Mammalian Endoplasmic Reticulum. *Cold Spring Harb. Perspect. Biol.* **2012**, *4*, a013219. [CrossRef]

114. Bellisola, G.; Fracasso, G.; Ippoliti, R.; Menestrina, G.; Rosén, A.; Soldà, S.; Udali, S.; Tomazzolli, R.; Tridente, G.; Colombatti, M. Reductive activation of ricin and ricin A-chain immunotoxins by protein disulfide isomerase and thioredoxin reductase. *Biochem. Pharmacol.* **2004**, *67*, 1721–1731. [CrossRef] [PubMed]

115. Arnér, E.S.; Holmgren, A. Physiological functions of thioredoxin and thioredoxin reductase. *Eur. J. Biochem.* **2000**, *267*, 6102–6109. [CrossRef] [PubMed]

116. Holmgren, A. Antioxidant function of thioredoxin and glutaredoxin systems. *Antioxid. Redox Signal.* **2000**, *2*, 811–820. [CrossRef] [PubMed]

117. Richardson, P.T.; Westby, M.; Roberts, L.M.; Gould, J.H.; Colman, A.; Lord, J.M. Recombinant proricin binds galactose but does not depurinate 28 S ribosomal RNA. *FEBS Lett.* **1989**, *255*, 15–20. [CrossRef]

118. Pasetto, M.; Barison, E.; Castagna, M.; Della Cristina, P.; Anselmi, C.; Colombatti, M. Reductive activation of type 2 ribosome-inactivating proteins is promoted by transmembrane thioredoxin-related protein. *J. Biol. Chem.* **2012**, *287*, 7367–7373. [CrossRef]

119. Argent, R.H.; Roberts, L.M.; Wales, R.; Robertus, J.D.; Lord, J.M. Introduction of a disulfide bond into ricin A chain decreases the cytotoxicity of the ricin holotoxin. *J. Biol. Chem.* **1994**, *269*, 26705–26710.

120. Argent, R.H.; Parrott, A.M.; Day, P.J.; Roberts, L.M.; Stockley, P.G.; Lord, J.M.; Radford, S.E. Ribosome-mediated folding of partially unfolded ricin A-chain. *J. Biol. Chem.* **2000**, *275*, 9263–9269. [CrossRef]

121. Ellgaard, L.; Helenius, A. Quality control in the endoplasmic reticulum. *Nat. Rev. Mol. Cell Biol.* **2003**, *4*, 181–191. [CrossRef]

122. Hebert, D.N.; Molinari, M. In and out of the ER: Protein folding, quality control, degradation, and related human diseases. *Physiol. Rev.* **2007**, *87*, 1377–1408. [CrossRef]

123. Braakman, I.; Hebert, D.N. Protein Folding in the Endoplasmic Reticulum. *Cold Spring Harb. Perspect. Biol.* **2013**, *5*, a013201. [CrossRef]

124. Karagöz, G.E.; Acosta-Alvear, D.; Walter, P. The Unfolded Protein Response: Detecting and Responding to Fluctuations in the Protein-Folding Capacity of the Endoplasmic Reticulum. *Cold Spring Harb. Perspect. Biol.* **2019**, a033886. [CrossRef] [PubMed]

125. Tsai, B.; Ye, Y.; Rapoport, T.A. Retro-translocation of proteins from the endoplasmic reticulum into the cytosol. *Nat. Rev. Mol. Cell Biol.* **2002**, *3*, 246–255. [CrossRef] [PubMed]

126. Gregers, T.F.; Skånland, S.S.; Wälchli, S.; Bakke, O.; Sandvig, K. BiP negatively affects ricin transport. *Toxins* **2013**, *5*, 969–982. [CrossRef] [PubMed]

127. Sun, M.; Kotler, J.L.M.; Liu, S.; Street, T.O. The ER chaperones BiP and Grp94 selectively associate when BiP is in the ADP conformation. *J. Biol. Chem.* **2019**. [CrossRef] [PubMed]

128. Spooner, R.A.; Hart, P.J.; Cook, J.P.; Pietroni, P.; Rogon, C.; Höhfeld, J.; Roberts, L.M.; Lord, J.M. Cytosolic chaperones influence the fate of a toxin dislocated from the endoplasmic reticulum. *Proc. Natl. Acad. Sci. USA* **2008**, *105*, 17408–17413. [CrossRef] [PubMed]

129. Pearse, B.R.; Hebert, D.N. Lectin chaperones help direct the maturation of glycoproteins in the endoplasmic reticulum. *Biochim. Biophys. Acta Mol. Cell Res.* **2010**, *1803*, 684–693. [CrossRef]

130. Kapoor, M.; Srinivas, H.; Kandiah, E.; Gemma, E.; Ellgaard, L.; Oscarson, S.; Helenius, A.; Surolia, A. Interactions of Substrate with Calreticulin, an Endoplasmic Reticulum Chaperone. *J. Biol. Chem.* **2003**, *278*, 6194–6200. [CrossRef]

131. Hosokawa, N.; Wada, I.; Hasegawa, K.; Yorihuzi, T.; Tremblay, L.O.; Herscovics, A.; Nagata, K. A novel ER alpha-mannosidase-like protein accelerates ER-associated degradation. *EMBO Rep.* **2001**, *2*, 415–422. [CrossRef]

132. Mast, S.W.; Diekman, K.; Karaveg, K.; Davis, A.; Sifers, R.N.; Moremen, K.W. Human EDEM2, a novel homolog of family 47 glycosidases, is involved in ER-associated degradation of glycoproteins. *Glycobiology* **2005**, *15*, 421–436. [CrossRef]

133. Hirao, K.; Natsuka, Y.; Tamura, T.; Wada, I.; Morito, D.; Natsuka, S.; Romero, P.; Sleno, B.; Tremblay, L.O.; Herscovics, A.; et al. EDEM3, a soluble EDEM homolog, enhances glycoprotein endoplasmic reticulum-associated degradation and mannose trimming. *J. Biol. Chem.* **2006**, *281*, 9650–9658. [CrossRef]

134. Olivari, S.; Molinari, M. Glycoprotein folding and the role of EDEM1, EDEM2 and EDEM3 in degradation of folding-defective glycoproteins. *FEBS Lett.* **2007**, *581*, 3658–3664. [CrossRef] [PubMed]

135. Slominska-Wojewodzka, M.; Gregers, T.F.; Wälchli, S.; Sandvig, K. EDEM Is Involved in Retrotranslocation of Ricin from the Endoplasmic Reticulum to the Cytosol. *Mol. Biol. Cell* **2006**, *17*, 1664–1675. [CrossRef] [PubMed]

136. Słomińska-Wojewódzka, M.; Pawlik, A.; Sokołowska, I.; Antoniewicz, J.; Węgrzyn, G.; Sandvig, K. The role of EDEM2 compared with EDEM1 in ricin transport from the endoplasmic reticulum to the cytosol. *Biochem. J.* **2014**, *457*, 485–496. [CrossRef] [PubMed]

137. Słomińska-Wojewódzka, M.; Sandvig, K. The Role of Lectin-Carbohydrate Interactions in the Regulation of ER-Associated Protein Degradation. *Molecules* **2015**, *20*, 9816–9846. [CrossRef] [PubMed]

138. Sominka, H.; Nowakowska-Gołacka, J.; Sowa-Rogozińska, N.; Słomińska-Wojewódzka, M. The role of EDEM3 in ricin cytotoxicity and its transport from the ER to the cytosol. Unpublished. Manuscript in preparation.

139. Mayerhofer, P.U.; Cook, J.P.; Wahlman, J.; Pinheiro, T.T.J.; Moore, K.A.H.; Lord, J.M.; Johnson, A.E.; Roberts, L.M. Ricin A Chain Insertion into Endoplasmic Reticulum Membranes Is Triggered by a Temperature Increase to 37 °C. *J. Biol. Chem.* **2009**, *284*, 10232–10242. [CrossRef] [PubMed]

140. Katzin, B.J.; Collins, E.J.; Robertus, J.D. Structure of ricin A-chain at 2.5 A. *Proteins* **1991**, *10*, 251–259. [CrossRef] [PubMed]

141. Yan, Q.; Li, X.-P.; Tumer, N.E. N-glycosylation does not affect the catalytic activity of ricin a chain but stimulates cytotoxicity by promoting its transport out of the endoplasmic reticulum. *Traffic* **2012**, *13*, 1508–1521. [CrossRef] [PubMed]

142. Leto, D.E.; Morgens, D.W.; Zhang, L.; Walczak, C.P.; Elias, J.E.; Bassik, M.C.; Kopito, R.R. Genome-wide CRISPR Analysis Identifies Substrate-Specific Conjugation Modules in ER-Associated Degradation. *Mol. Cell* **2019**, *73*, 377–389. [CrossRef]

143. Pilon, M.; Schekman, R.; Römisch, K. Sec61p mediates export of a misfolded secretory protein from the endoplasmic reticulum to the cytosol for degradation. *EMBO J.* **1997**, *16*, 4540–4548. [CrossRef]

144. Schäfer, A.; Wolf, D.H. Sec61p is part of the endoplasmic reticulum-associated degradation machinery. *EMBO J.* **2009**, *28*, 2874–2884. [CrossRef]

145. Tretter, T.; Pereira, F.P.; Ulucan, O.; Helms, V.; Allan, S.; Kalies, K.-U.; Römisch, K. ERAD and protein import defects in a sec61 mutant lacking ER-lumenal loop 7. *BMC Cell Biol.* **2013**, *14*, 56. [CrossRef] [PubMed]

146. Kaiser, M.-L.; Römisch, K. Proteasome 19S RP Binding to the Sec61 Channel Plays a Key Role in ERAD. *PLoS ONE* **2015**, *10*, e0117260. [CrossRef] [PubMed]

147. Römisch, K. A Case for Sec61 Channel Involvement in ERAD. *Trends Biochem. Sci.* **2017**, *42*, 171–179. [CrossRef] [PubMed]

148. Lilley, B.N.; Ploegh, H.L. A membrane protein required for dislocation of misfolded proteins from the ER. *Nature* **2004**, *429*, 834–840. [CrossRef] [PubMed]

149. Ye, Y.; Shibata, Y.; Yun, C.; Ron, D.; Rapoport, T.A. A membrane protein complex mediates retro-translocation from the ER lumen into the cytosol. *Nature* **2004**, *429*, 841–847. [CrossRef] [PubMed]

150. Lilley, B.N.; Ploegh, H.L. Multiprotein complexes that link dislocation, ubiquitination, and extraction of misfolded proteins from the endoplasmic reticulum membrane. *Proc. Natl. Acad. Sci. USA* **2005**, *102*, 14296–14301. [CrossRef] [PubMed]

151. Oda, Y.; Okada, T.; Yoshida, H.; Kaufman, R.J.; Nagata, K.; Mori, K. Derlin-2 and Derlin-3 are regulated by the mammalian unfolded protein response and are required for ER-associated degradation. *J. Cell Biol.* **2006**, *172*, 383–393. [CrossRef]

152. You, H.; Ge, Y.; Zhang, J.; Cao, Y.; Xing, J.; Su, D.; Huang, Y.; Li, M.; Qu, S.; Sun, F.; et al. Derlin-1 promotes ubiquitylation and degradation of the epithelial Na+ channel, ENaC. *J. Cell Sci.* **2017**, *130*, 1027–1036.

153. Mehnert, M.; Sommer, T.; Jarosch, E. Der1 promotes movement of misfolded proteins through the endoplasmic reticulum membrane. *Nat. Cell Biol.* **2014**, *16*, 77–86. [CrossRef]

154. Kikkert, M.; Doolman, R.; Dai, M.; Avner, R.; Hassink, G.; van Voorden, S.; Thanedar, S.; Roitelman, J.; Chau, V.; Wiertz, E. Human HRD1 is an E3 ubiquitin ligase involved in degradation of proteins from the endoplasmic reticulum. *J. Biol. Chem.* **2004**, *279*, 3525–3534. [CrossRef]

155. Carvalho, P.; Stanley, A.M.; Rapoport, T.A. Retrotranslocation of a misfolded luminal ER protein by the ubiquitin-ligase Hrd1p. *Cell* **2010**, *143*, 579–591. [CrossRef]

156. Stein, A.; Ruggiano, A.; Carvalho, P.; Rapoport, T.A. Key steps in ERAD of luminal ER proteins reconstituted with purified components. *Cell* **2014**, *158*, 1375–1388. [CrossRef] [PubMed]

157. Baldridge, R.D.; Rapoport, T.A. Autoubiquitination of the Hrd1 Ligase Triggers Protein Retrotranslocation in ERAD. *Cell* **2016**, *166*, 394–407. [CrossRef] [PubMed]

158. Schoebel, S.; Mi, W.; Stein, A.; Ovchinnikov, S.; Pavlovicz, R.; DiMaio, F.; Baker, D.; Chambers, M.G.; Su, H.; Li, D.; et al. Cryo-EM structure of the protein-conducting ERAD channel Hrd1 in complex with Hrd3. *Nature* **2017**, *548*, 352–355. [CrossRef] [PubMed]

159. Hwang, J.; Qi, L. Quality Control in the Endoplasmic Reticulum: Crosstalk between ERAD and UPR pathways. *Trends Biochem. Sci.* **2018**, *43*, 593–605. [CrossRef]

160. Zheng, N.; Shabek, N. Ubiquitin Ligases: Structure, Function, and Regulation. *Annu. Rev. Biochem.* **2017**, *86*, 129–157. [CrossRef]

161. Sowa-Rogozińska, N.; Słomińska-Wojewódzka, M. The role of Sec61 in ricin transport from the ER to the cytosol. Unpublished. Manuscript in preparation.

162. Sowa-Rogozińska, N.; Sominka, H.; Słomińska-Wojewódzka, M. The role of Derlin proteins in ricin transport from the ER to the cytosol. Unpublished. Manuscript in preparation.

163. Dang, H.; Klokk, T.I.; Schaheen, B.; McLaughlin, B.M.; Thomas, A.J.; Durns, T.A.; Bitler, B.G.; Sandvig, K.; Fares, H. Derlin-dependent retrograde transport from endosomes to the Golgi apparatus. *Traffic* **2011**, *12*, 1417–1431. [CrossRef] [PubMed]

164. Li, S.; Spooner, R.A.; Allen, S.C.H.; Guise, C.P.; Ladds, G.; Schnöder, T.; Schmitt, M.J.; Lord, J.M.; Roberts, L.M. Folding-competent and folding-defective forms of ricin A chain have different fates after retrotranslocation from the endoplasmic reticulum. *Mol. Biol. Cell* **2010**, *21*, 2543–2554. [CrossRef] [PubMed]

165. Eshraghi, A.; Dixon, S.D.; Tamilselvam, B.; Kim, E.J.-K.; Gargi, A.; Kulik, J.C.; Damoiseaux, R.; Blanke, S.R.; Bradley, K.A. Cytolethal distending toxins require components of the ER-associated degradation pathway for host cell entry. *PLoS Pathog.* **2014**, *10*, e1004295. [CrossRef]

166. Pietroni, P.; Vasisht, N.; Cook, J.P.; Roberts, D.M.; Lord, J.M.; Hartmann-Petersen, R.; Roberts, L.M.; Spooner, R.A. The proteasome cap RPT5/Rpt5p subunit prevents aggregation of unfolded ricin A chain. *Biochem. J.* **2013**, *453*, 435–445. [CrossRef]

167. Di Cola, A.; Frigerio, L.; Lord, J.M.; Roberts, L.M.; Ceriotti, A. Endoplasmic Reticulum-Associated Degradation of Ricin A Chain Has Unique and Plant-Specific Features. *Plant Physiol.* **2005**, *137*, 287–296. [CrossRef]

168. Deeks, E.D.; Cook, J.P.; Day, P.J.; Smith, D.C.; Roberts, L.M.; Lord, J.M. The low lysine content of ricin A chain reduces the risk of proteolytic degradation after translocation from the endoplasmic reticulum to the cytosol. *Biochemistry* **2002**, *41*, 3405–3413. [CrossRef] [PubMed]

169. Abujarour, R.J.; Dalal, S.; Hanson, P.I.; Draper, R.K. p97 Is in a complex with cholera toxin and influences the transport of cholera toxin and related toxins to the cytoplasm. *J. Biol. Chem.* **2005**, *280*, 15865–15871. [CrossRef] [PubMed]

170. Thrower, J.S.; Hoffman, L.; Rechsteiner, M.; Pickart, C.M. Recognition of the polyubiquitin proteolytic signal. *EMBO J.* **2000**, *19*, 94–102. [CrossRef] [PubMed]

171. Rodighiero, C.; Tsai, B.; Rapoport, T.A.; Lencer, W.I. Role of ubiquitination in retro-translocation of cholera toxin and escape of cytosolic degradation. *EMBO Rep.* **2002**, *3*, 1222–1227. [CrossRef] [PubMed]

172. Lipson, C.; Alalouf, G.; Bajorek, M.; Rabinovich, E.; Atir-Lande, A.; Glickman, M.; Bar-Nun, S. A proteasomal ATPase contributes to dislocation of endoplasmic reticulum-associated degradation (ERAD) substrates. *J. Biol. Chem.* **2008**, *283*, 7166–7175. [CrossRef] [PubMed]

173. Odunuga, O.O.; Longshaw, V.M.; Blatch, G.L. Hop: More than an Hsp70/Hsp90 adaptor protein. *Bioessays* **2004**, *26*, 1058–1068. [CrossRef] [PubMed]

174. Li, X.-P.; Grela, P.; Krokowski, D.; Tchórzewski, M.; Tumer, N.E. Pentameric organization of the ribosomal stalk accelerates recruitment of ricin a chain to the ribosome for depurination. *J. Biol. Chem.* **2010**, *285*, 41463–41471. [CrossRef] [PubMed]

175. Larsson, S.L.; Sloma, M.S.; Nygård, O. Conformational changes in the structure of domains II and V of 28S rRNA in ribosomes treated with the translational inhibitors ricin or alpha-sarcin. *Biochim. Biophys. Acta* **2002**, *1577*, 53–62. [CrossRef]

176. Endo, Y.; Tsurugi, K. The RNA N-glycosidase activity of ricin A-chain. The characteristics of the enzymatic activity of ricin A-chain with ribosomes and with rRNA. *J. Biol. Chem.* **1988**, *263*, 8735–8739.

177. May, K.L.; Yan, Q.; Tumer, N.E. Targeting ricin to the ribosome. *Toxicon* **2013**, *69*, 143–151. [CrossRef] [PubMed]

178. Chiou, J.-C.; Li, X.-P.; Remacha, M.; Ballesta, J.P.G.; Tumer, N.E. The ribosomal stalk is required for ribosome binding, depurination of the rRNA and cytotoxicity of ricin A chain in Saccharomyces cerevisiae. *Mol. Microbiol.* **2008**, *70*, 1441–1452. [CrossRef] [PubMed]

179. May, K.L.; Li, X.-P.; Martínez-Azorín, F.; Ballesta, J.P.G.; Grela, P.; Tchórzewski, M.; Tumer, N.E. The P1/P2 proteins of the human ribosomal stalk are required for ribosome binding and depurination by ricin in human cells. *FEBS J.* **2012**, *279*, 3925–3936. [CrossRef] [PubMed]

180. Grela, P.; Sawa-Makarska, J.; Gordiyenko, Y.; Robinson, C.V.; Grankowski, N.; Tchórzewski, M. Structural properties of the human acidic ribosomal P proteins forming the P1-P2 heterocomplex. *J. Biochem.* **2008**, *143*, 169–177. [CrossRef] [PubMed]

181. Lee, K.-M.; Yu, C.W.-H.; Chiu, T.Y.-H.; Sze, K.-H.; Shaw, P.-C.; Wong, K.-B. Solution structure of the dimerization domain of the eukaryotic stalk P1/P2 complex reveals the structural organization of eukaryotic stalk complex. *Nucleic Acids Res.* **2012**, *40*, 3172–3182. [CrossRef] [PubMed]

182. Grela, P.; Li, X.-P.; Horbowicz, P.; Dźwierzyńska, M.; Tchórzewski, M.; Tumer, N.E. Human ribosomal P1-P2 heterodimer represents an optimal docking site for ricin A chain with a prominent role for P1 C-terminus. *Sci. Rep.* **2017**, *7*, 5608. [CrossRef] [PubMed]

183. Grela, P.; Li, X.-P.; Tchórzewski, M.; Tumer, N.E. Functional divergence between the two P1-P2 stalk dimers on the ribosome in their interaction with ricin A chain. *Biochem. J.* **2014**, *460*, 59–67. [CrossRef]

184. Grela, P.; Bernadó, P.; Svergun, D.; Kwiatowski, J.; Abramczyk, D.; Grankowski, N.; Tchórzewski, M. Structural relationships among the ribosomal stalk proteins from the three domains of life. *J. Mol. Evol.* **2008**, *67*, 154–167. [CrossRef] [PubMed]

185. Li, X.-P.; Chiou, J.-C.; Remacha, M.; Ballesta, J.P.G.; Tumer, N.E. A two-step binding model proposed for the electrostatic interactions of ricin A chain with ribosomes. *Biochemistry* **2009**, *48*, 3853–3863. [CrossRef] [PubMed]

186. Tumer, N.E.; Li, X.-P. Interaction of ricin and Shiga toxins with ribosomes. *Curr. Top. Microbiol. Immunol.* **2012**, *357*, 1–18.

187. Fan, X.; Zhu, Y.; Wang, C.; Niu, L.; Teng, M.; Li, X. Structural insights into the interaction of the ribosomal P stalk protein P2 with a type II ribosome-inactivating protein ricin. *Sci. Rep.* **2016**, *6*, 37803. [CrossRef] [PubMed]

188. Shi, W.-W.; Tang, Y.-S.; Sze, S.-Y.; Zhu, Z.-N.; Wong, K.-B.; Shaw, P.-C. Crystal Structure of Ribosome-Inactivating Protein Ricin A Chain in Complex with the C-Terminal Peptide of the Ribosomal Stalk Protein P2. *Toxins* **2016**, *8*, 296. [CrossRef] [PubMed]

189. Watanabe, K.; Dansako, H.; Asada, N.; Sakai, M.; Funatsu, G. Effects of chemical modification of arginine residues outside the active site cleft of ricin A-chain on its RNA N-glycosidase activity for ribosomes. *Biosci. Biotechnol. Biochem.* **1994**, *58*, 716–721. [CrossRef] [PubMed]

190. Marsden, C.J.; Fülöp, V.; Day, P.J.; Lord, J.M. The effect of mutations surrounding and within the active site on the catalytic activity of ricin A chain. *Eur. J. Biochem.* **2004**, *271*, 153–162. [CrossRef] [PubMed]

191. Zhou, Y.; Li, X.-P.; Chen, B.Y.; Tumer, N.E. Ricin uses arginine 235 as an anchor residue to bind to P-proteins of the ribosomal stalk. *Sci. Rep.* **2017**, *7*, 42912. [CrossRef] [PubMed]

192. Li, X.-P.; Kahn, P.C.; Kahn, J.N.; Grela, P.; Tumer, N.E. Arginine residues on the opposite side of the active site stimulate the catalysis of ribosome depurination by ricin A chain by interacting with the P-protein stalk. *J. Biol. Chem.* **2013**, *288*, 30270–30284. [CrossRef] [PubMed]

193. Li, X.-P.; Tumer, N.E. Differences in Ribosome Binding and Sarcin/Ricin Loop Depurination by Shiga and Ricin Holotoxins. *Toxins* **2017**, *9*, 133. [CrossRef] [PubMed]

194. Jetzt, A.E.; Li, X.-P.; Tumer, N.E.; Cohick, W.S. Toxicity of ricin A chain is reduced in mammalian cells by inhibiting its interaction with the ribosome. *Toxicol. Appl. Pharmacol.* **2016**, *310*, 120–128. [CrossRef]

195. Hedblom, M.L.; Cawley, D.B.; Boguslawski, S.; Houston, L.L. Binding of ricin A chain to rat liver ribosomes: Relationship to ribosome inactivation. *J. Supramol. Struct.* **1978**, *9*, 253–268. [CrossRef]

196. Honjo, E.; Watanabe, K.; Tsukamoto, T. Real-time kinetic analyses of the interaction of ricin toxin A-chain with ribosomes prove a conformational change involved in complex formation. *J. Biochem.* **2002**, *131*, 267–275. [CrossRef]

197. Dai, J.; Zhao, L.; Yang, H.; Guo, H.; Fan, K.; Wang, H.; Qian, W.; Zhang, D.; Li, B.; Wang, H.; et al. Identification of a novel functional domain of ricin responsible for its potent toxicity. *J. Biol. Chem.* **2011**, *286*, 12166–12171. [CrossRef] [PubMed]

198. Morris, K.N.; Wool, I.G. Determination by systematic deletion of the amino acids essential for catalysis by ricin A chain. *Proc. Natl. Acad. Sci. USA* **1992**, *89*, 4869–4873. [CrossRef] [PubMed]

199. Day, P.J.; Ernst, S.R.; Frankel, A.E.; Monzingo, A.F.; Pascal, J.M.; Molina-Svinth, M.C.; Robertus, J.D. Structure and activity of an active site substitution of ricin A chain. *Biochemistry* **1996**, *35*, 11098–11103. [CrossRef] [PubMed]

200. Li, X.-P.; Kahn, J.N.; Tumer, N.E. Peptide Mimics of the Ribosomal P Stalk Inhibit the Activity of Ricin A Chain by Preventing Ribosome Binding. *Toxins* **2018**, *10*, 371. [CrossRef]

201. Lewis, J.L.; Shields, K.A.; Chong, D.C. Detection and quantification of ricin-mediated 28S ribosomal depurination by digital droplet PCR. *Anal. Biochem.* **2018**, *563*, 15–19. [CrossRef]

202. Falach, R.; Sapoznikov, A.; Gal, Y.; Israeli, O.; Leitner, M.; Seliger, N.; Ehrlich, S.; Kronman, C.; Sabo, T. Quantitative profiling of the in vivo enzymatic activity of ricin reveals disparate depurination of different pulmonary cell types. *Toxicol. Lett.* **2016**, *258*, 11–19. [CrossRef]

203. Meneguelli de Souza, L.C.; de Carvalho, L.P.; Araújo, J.S.; de Melo, E.J.T.; Machado, O.L.T. Cell toxicity by ricin and elucidation of mechanism of Ricin inactivation. *Int. J. Biol. Macromol.* **2018**, *113*, 821–828. [CrossRef]

204. Sandvig, K.; van Deurs, B. Toxin-induced cell lysis: Protection by 3-methyladenine and cycloheximide. *Exp. Cell Res.* **1992**, *200*, 253–262. [CrossRef]

205. Bagaria, S.; Karande, A. Abrin and Immunoneutralization: A Review. *Toxinology* **2014**. [CrossRef]

206. Bingen, A.; Creppy, E.E.; Gut, J.P.; Dirheimer, G.; Kirn, A. The Kupffer cell is the first target in ricin-induced hepatitis. *J. Submicrosc. Cytol.* **1987**, *19*, 247–256.

207. Leek, M.D.; Griffiths, G.D.; Green, M.A. Intestinal pathology following intramuscular ricin poisoning. *J. Pathol.* **1989**, *159*, 329–334. [CrossRef] [PubMed]

208. Wilhelmsen, C.L.; Pitt, M.L. Lesions of acute inhaled lethal ricin intoxication in rhesus monkeys. *Vet. Pathol.* **1996**, *33*, 296–302. [CrossRef] [PubMed]

209. Soler-Rodríguez, A.M.; Ghetie, M.A.; Oppenheimer-Marks, N.; Uhr, J.W.; Vitetta, E.S. Ricin A-chain and ricin A-chain immunotoxins rapidly damage human endothelial cells: Implications for vascular leak syndrome. *Exp. Cell Res.* **1993**, *206*, 227–234. [CrossRef] [PubMed]

210. Kochi, S.K.; Collier, R.J. DNA fragmentation and cytolysis in U937 cells treated with diphtheria toxin or other inhibitors of protein synthesis. *Exp. Cell Res.* **1993**, *208*, 296–302. [CrossRef] [PubMed]

211. Peumans, W.J.; Hao, Q.; Van Damme, E.J. Ribosome-inactivating proteins from plants: More than RNA N-glycosidases? *FASEB J.* **2001**, *15*, 1493–1506. [CrossRef] [PubMed]

212. Brigotti, M.; Alfieri, R.; Sestili, P.; Bonelli, M.; Petronini, P.G.; Guidarelli, A.; Barbieri, L.; Stirpe, F.; Sperti, S. Damage to nuclear DNA induced by Shiga toxin 1 and ricin in human endothelial cells. *FASEB J.* **2002**, *16*, 365–372. [CrossRef] [PubMed]

213. Li, M.; Pestka, J.J. Comparative induction of 28S ribosomal RNA cleavage by ricin and the trichothecenes deoxynivalenol and T-2 toxin in the macrophage. *Toxicol. Sci.* **2008**, *105*, 67–78. [CrossRef] [PubMed]

214. Gray, J.S.; Bae, H.K.; Li, J.C.B.; Lau, A.S.; Pestka, J.J. Double-stranded RNA-activated protein kinase mediates induction of interleukin-8 expression by deoxynivalenol, Shiga toxin 1, and ricin in monocytes. *Toxicol. Sci.* **2008**, *105*, 322–330. [CrossRef] [PubMed]

215. Tesh, V.L. The induction of apoptosis by Shiga toxins and ricin. *Curr. Top. Microbiol. Immunol.* **2012**, *357*, 137–178. [PubMed]

216. Komatsu, N.; Oda, T.; Muramatsu, T. Involvement of both caspase-like proteases and serine proteases in apoptotic cell death induced by ricin, modeccin, diphtheria toxin, and pseudomonas toxin. *J. Biochem.* **1998**, *124*, 1038–1044. [CrossRef] [PubMed]

217. Williams, J.M.; Lea, N.; Lord, J.M.; Roberts, L.M.; Milford, D.V.; Taylor, C.M. Comparison of ribosome-inactivating proteins in the induction of apoptosis. *Toxicol. Lett.* **1997**, *91*, 121–127. [CrossRef]

218. Rao, P.V.L.; Jayaraj, R.; Bhaskar, A.S.B.; Kumar, O.; Bhattacharya, R.; Saxena, P.; Dash, P.K.; Vijayaraghavan, R. Mechanism of ricin-induced apoptosis in human cervical cancer cells. *Biochem. Pharmacol.* **2005**, *69*, 855–865. [CrossRef] [PubMed]

219. Jetzt, A.E.; Cheng, J.-S.; Tumer, N.E.; Cohick, W.S. Ricin A-chain requires c-Jun N-terminal kinase to induce apoptosis in nontransformed epithelial cells. *Int. J. Biochem. Cell Biol.* **2009**, *41*, 2503–2510. [CrossRef] [PubMed]

220. Komatsu, N.; Nakagawa, M.; Oda, T.; Muramatsu, T. Depletion of intracellular NAD(+) and ATP levels during ricin-induced apoptosis through the specific ribosomal inactivation results in the cytolysis of U937 cells. *J. Biochem.* **2000**, *128*, 463–470. [CrossRef] [PubMed]

221. Wu, Y.-H.; Shih, S.-F.; Lin, J.-Y. Ricin Triggers Apoptotic Morphological Changes through Caspase-3 Cleavage of BAT3. *J. Biol. Chem.* **2004**, *279*, 19264–19275. [CrossRef] [PubMed]

222. Sha, O.; Yew, D.T.W.; Ng, T.B.; Yuan, L.; Kwong, W.H. Different in vitro toxicities of structurally similar type I ribosome-inactivating proteins (RIPs). *Toxicol. In Vitro* **2010**, *24*, 1176–1182. [CrossRef] [PubMed]

223. Brinkmann, U.; Mansfield, E.; Pastan, I. Effects of BCL-2 overexpression on the sensitivity of MCF-7 breast cancer cells to ricin, diphtheria and Pseudomonas toxin and immunotoxins. *Apoptosis* **1997**, *2*, 192–198. [CrossRef] [PubMed]

224. Hu, R.; Zhai, Q.; Liu, W.; Liu, X. An insight into the mechanism of cytotoxicity of ricin to hepatoma cell: Roles of Bcl-2 family proteins, caspases, Ca(2+)-dependent proteases and protein kinase C. *J. Cell. Biochem.* **2001**, *81*, 583–593. [CrossRef] [PubMed]

225. Tamura, T.; Oda, T.; Muramatsu, T. Resistance against ricin-induced apoptosis in a brefeldin A-resistant mutant cell line (BER-40) of Vero cells. *J. Biochem.* **2002**, *132*, 441–449. [CrossRef] [PubMed]

226. Iordanov, M.S.; Pribnow, D.; Magun, J.L.; Dinh, T.H.; Pearson, J.A.; Chen, S.L.; Magun, B.E. Ribotoxic stress response: Activation of the stress-activated protein kinase JNK1 by inhibitors of the peptidyl transferase reaction and by sequence-specific RNA damage to the alpha-sarcin/ricin loop in the 28S rRNA. *Mol. Cell Biol.* **1997**, *17*, 3373–3381. [CrossRef]

227. Korcheva, V.; Wong, J.; Corless, C.; Iordanov, M.; Magun, B. Administration of ricin induces a severe inflammatory response via nonredundant stimulation of ERK, JNK, and P38 MAPK and provides a mouse model of hemolytic uremic syndrome. *Am. J. Pathol.* **2005**, *166*, 323–339. [CrossRef]

228. Gonzalez, T.V.; Farrant, S.A.; Mantis, N.J. Ricin induces IL-8 secretion from human monocyte/macrophages by activating the p38 MAP kinase pathway. *Mol. Immunol.* **2006**, *43*, 1920–1923. [CrossRef] [PubMed]

229. Korcheva, V.; Wong, J.; Lindauer, M.; Jacoby, D.B.; Iordanov, M.S.; Magun, B. Role of Apoptotic Signaling Pathways in Regulation of Inflammatory Responses to Ricin in Primary Murine Macrophages. *Mol. Immunol.* **2007**, *44*, 2761–2771. [CrossRef] [PubMed]

230. Wong, J.; Korcheva, V.; Jacoby, D.B.; Magun, B.E. Proinflammatory responses of human airway cells to ricin involve stress-activated protein kinases and NF-kappaB. *Am. J. Physiol. Lung Cell. Mol. Physiol.* **2007**, *293*, L1385–L1394. [CrossRef] [PubMed]

231. Yamasaki, C.; Nishikawa, K.; Zeng, X.-T.; Katayama, Y.; Natori, Y.; Komatsu, N.; Oda, T.; Natori, Y. Induction of cytokines by toxins that have an identical RNA N-glycosidase activity: Shiga toxin, ricin, and modeccin. *Biochim. Biophys. Acta* **2004**, *1671*, 44–50. [CrossRef] [PubMed]

232. Lindauer, M.L.; Wong, J.; Iwakura, Y.; Magun, B.E. Pulmonary inflammation triggered by ricin toxin requires macrophages and IL-1 signaling. *J. Immunol.* **2009**, *183*, 1419–1426. [CrossRef] [PubMed]

233. Lindauer, M.; Wong, J.; Magun, B. Ricin Toxin Activates the NALP3 Inflammasome. *Toxins* **2010**, *2*, 1500–1514. [CrossRef]

234. Gal, Y.; Mazor, O.; Falach, R.; Sapoznikov, A.; Kronman, C.; Sabo, T. Treatments for Pulmonary Ricin Intoxication: Current Aspects and Future Prospects. *Toxins* **2017**, *9*, 311. [CrossRef]

235. Jandhyala, D.M.; Ahluwalia, A.; Obrig, T.; Thorpe, C.M. ZAK: A MAP3Kinase that transduces Shiga toxin- and ricin-induced proinflammatory cytokine expression. *Cell. Microbiol.* **2008**, *10*, 1468–1477. [CrossRef]

236. Higuchi, S.; Tamura, T.; Oda, T. Cross-talk between the pathways leading to the induction of apoptosis and the secretion of tumor necrosis factor-alpha in ricin-treated RAW 264.7 cells. *J. Biochem.* **2003**, *134*, 927–933. [CrossRef]

237. Wang, C.-T.; Jetzt, A.E.; Cheng, J.-S.; Cohick, W.S. Inhibition of the Unfolded Protein Response by Ricin A-Chain Enhances Its Cytotoxicity in Mammalian Cells. *Toxins* **2011**, *3*, 453–468. [CrossRef] [PubMed]

238. Horrix, C.; Raviv, Z.; Flescher, E.; Voss, C.; Berger, M.R. Plant ribosome-inactivating proteins type II induce the unfolded protein response in human cancer cells. *Cell. Mol. Life Sci.* **2011**, *68*, 1269–1281. [CrossRef] [PubMed]

239. Barbieri, L.; Brigotti, M.; Perocco, P.; Carnicelli, D.; Ciani, M.; Mercatali, L.; Stirpe, F. Ribosome-inactivating proteins depurinate poly(ADP-ribosyl) ated poly(ADP-ribose) polymerase and have transforming activity for 3T3 fibroblasts. *FEBS Lett.* **2003**, *538*, 178–182. [CrossRef]

240. Oda, T.; Iwaoka, J.; Komatsu, N.; Muramatsu, T. Involvement of N-acetylcysteine-sensitive pathways in ricin-induced apoptotic cell death in U937 cells. *Biosci. Biotechnol. Biochem.* **1999**, *63*, 341–348. [CrossRef] [PubMed]

241. Oda, T.; Komatsu, N.; Muramatsu, T. Inhibitory effect of dideoxyforskolin on cell death induced by ricin, modeccin, diphtheria toxin, and Pseudomonas toxin in MDCK cells. *Cell Struct. Funct.* **1997**, *22*, 545–554. [CrossRef] [PubMed]

242. Tamura, T.; Sadakata, N.; Oda, T.; Muramatsu, T. Role of zinc ions in ricin-induced apoptosis in U937 cells. *Toxicol. Lett.* **2002**, *132*, 141–151. [CrossRef]

243. Authier, F.; Djavaheri-Mergny, M.; Lorin, S.; Frénoy, J.-P.; Desbuquois, B. Fate and action of ricin in rat liver in vivo: Translocation of endocytosed ricin into cytosol and induction of intrinsic apoptosis by ricin B-chain. *Cell. Microbiol.* **2016**, *18*, 1800–1814. [CrossRef] [PubMed]

244. Elmore, S. Apoptosis: A review of programmed cell death. *Toxicol. Pathol.* **2007**, *35*, 495–516. [CrossRef]

245. Nagata, S. Apoptosis and Clearance of Apoptotic Cells. *Annu. Rev. Immunol.* **2018**, *36*, 489–517. [CrossRef] [PubMed]

246. Boulares, A.H.; Yakovlev, A.G.; Ivanova, V.; Stoica, B.A.; Wang, G.; Iyer, S.; Smulson, M. Role of poly(ADP-ribose) polymerase (PARP) cleavage in apoptosis. Caspase 3-resistant PARP mutant increases rates of apoptosis in transfected cells. *J. Biol. Chem.* **1999**, *274*, 22932–22940. [CrossRef]

247. Kitazumi, I.; Tsukahara, M. Regulation of DNA fragmentation: The role of caspases and phosphorylation. *FEBS J.* **2011**, *278*, 427–441. [CrossRef] [PubMed]

248. Desmots, F.; Russell, H.R.; Michel, D.; McKinnon, P.J. Scythe regulates apoptosis-inducing factor stability during endoplasmic reticulum stress-induced apoptosis. *J. Biol. Chem.* **2008**, *283*, 3264–3271. [CrossRef] [PubMed]

249. Czabotar, P.E.; Lessene, G.; Strasser, A.; Adams, J.M. Control of apoptosis by the BCL-2 protein family: Implications for physiology and therapy. *Nat. Rev. Mol. Cell Biol.* **2014**, *15*, 49–63. [CrossRef]

250. Lin, A.; Minden, A.; Martinetto, H.; Claret, F.X.; Lange-Carter, C.; Mercurio, F.; Johnson, G.L.; Karin, M. Identification of a dual specificity kinase that activates the Jun kinases and p38-Mpk2. *Science* **1995**, *268*, 286–290. [CrossRef] [PubMed]

251. Cargnello, M.; Roux, P.P. Activation and function of the MAPKs and their substrates, the MAPK-activated protein kinases. *Microbiol. Mol. Biol. Rev.* **2011**, *75*, 50–83. [CrossRef] [PubMed]

252. Liu, H.; Ma, Y.; Pagliari, L.J.; Perlman, H.; Yu, C.; Lin, A.; Pope, R.M. TNF-alpha-induced apoptosis of macrophages following inhibition of NF-kappa B: A central role for disruption of mitochondria. *J. Immunol.* **2004**, *172*, 1907–1915. [CrossRef] [PubMed]

253. Papa, S.; Bubici, C.; Zazzeroni, F.; Pham, C.G.; Kuntzen, C.; Knabb, J.R.; Dean, K.; Franzoso, G. The NF-kappaB-mediated control of the JNK cascade in the antagonism of programmed cell death in health and disease. *Cell Death Differ.* **2006**, *13*, 712–729. [CrossRef] [PubMed]

254. Bernales, S.; Papa, F.R.; Walter, P. Intracellular signaling by the unfolded protein response. *Annu. Rev. Cell Dev. Biol.* **2006**, *22*, 487–508. [CrossRef]

255. Adams, C.J.; Kopp, M.C.; Larburu, N.; Nowak, P.R.; Ali, M.M.U. Structure and Molecular Mechanism of ER Stress Signaling by the Unfolded Protein Response Signal Activator IRE1. *Front. Mol. Biosci.* **2019**, *6*, 11. [CrossRef]

256. Guerra-Moreno, A.; Ang, J.; Welsch, H.; Jochem, M.; Hanna, J. Regulation of the unfolded protein response in yeast by oxidative stress. *FEBS Lett.* **2019**, *593*, 1080–1088. [CrossRef]

257. Barbieri, L.; Valbonesi, P.; Bonora, E.; Gorini, P.; Bolognesi, A.; Stirpe, F. Polynucleotide: Adenosine glycosidase activity of ribosome-inactivating proteins: Effect on DNA, RNA and poly(A). *Nucleic Acids Res.* **1997**, *25*, 518–522. [CrossRef] [PubMed]

258. D'Amours, D.; Desnoyers, S.; D'Silva, I.; Poirier, G.G. Poly(ADP-ribosyl) ation reactions in the regulation of nuclear functions. *Biochem. J.* **1999**, *342 Pt 2*, 249–268. [CrossRef]

259. Sestili, P.; Alfieri, R.; Carnicelli, D.; Martinelli, C.; Barbieri, L.; Stirpe, F.; Bonelli, M.; Petronini, P.G.; Brigotti, M. Shiga toxin 1 and ricin inhibit the repair of H_2O_2-induced DNA single strand breaks in cultured mammalian cells. *DNA Repair* **2005**, *4*, 271–277. [CrossRef] [PubMed]

260. Circu, M.L.; Aw, T.Y. Reactive oxygen species, cellular redox systems, and apoptosis. *Free Radic. Biol. Med.* **2010**, *48*, 749–762. [CrossRef] [PubMed]

261. Zhang, C.; Gong, Y.; Ma, H.; An, C.; Chen, D.; Chen, Z.L. Reactive oxygen species involved in trichosanthin-induced apoptosis of human choriocarcinoma cells. *Biochem. J.* **2001**, *355*, 653–661. [CrossRef] [PubMed]

262. Ramsden, C.S.; Drayson, M.T.; Bell, E.B. The toxicity, distribution and excretion of ricin holotoxin in rats. *Toxicology* **1989**, *55*, 161–171. [CrossRef]

263. Xu, N.; Yuan, H.; Liu, W.; Li, S.; Liu, Y.; Wan, J.; Li, X.; Zhang, R.; Chang, Y. Activation of RAW264.7 mouse macrophage cells in vitro through treatment with recombinant ricin toxin-binding subunit B: Involvement of protein tyrosine, NF-κB and JAK-STAT kinase signaling pathways. *Int. J. Mol. Med.* **2013**, *32*, 729–735. [CrossRef] [PubMed]

264. Silverstein, A.M. The collected papers of Paul Ehrlich: Why was volume 4 never published? *Bull. Hist. Med.* **2002**, *76*, 335–339. [CrossRef] [PubMed]

265. Vitetta, E.S.; Thorpe, P.E.; Uhr, J.W. Immunotoxins: Magic bullets or misguided missiles? *Immunol. Today* **1993**, *14*, 252–259. [CrossRef]

266. Brinkmann, U.; Pastan, I. Immunotoxins against cancer. *Biochim. Biophys. Acta* **1994**, *1198*, 27–45. [CrossRef]

267. Weidle, U.H.; Tiefenthaler, G.; Schiller, C.; Weiss, E.H.; Georges, G.; Brinkmann, U. Prospects of bacterial and plant protein-based immunotoxins for treatment of cancer. *Cancer Genomics Proteomics* **2014**, *11*, 25–38. [PubMed]

268. Munir, I.; Naseer, R.; Saleem, M.; Mahmood, A.; Sultana, A. Immunotoxins, an Advance tool for Cancer Treatment: Review and update. *Acta Pol. Pharm. Drug Res.* **2018**, *75*, 1267–1277. [CrossRef]

269. Polito, L.; Djemil, A.; Bortolotti, M. Plant Toxin-Based Immunotoxins for Cancer Therapy: A Short Overview. *Biomedicines* **2016**, *4*, 12. [CrossRef] [PubMed]

270. Lambert, J.M.; Goldmacher, V.S.; Collinson, A.R.; Nadler, L.M.; Blättler, W.A. An immunotoxin prepared with blocked ricin: A natural plant toxin adapted for therapeutic use. *Cancer Res.* **1991**, *51*, 6236–6242.

271. Krolick, K.A.; Villemez, C.; Isakson, P.; Uhr, J.W.; Vitetta, E.S. Selective killing of normal or neoplastic B cells by antibodies coupled to the A chain of ricin. *Proc. Natl. Acad. Sci. USA* **1980**, *77*, 5419–5423. [CrossRef] [PubMed]

272. Bourrie, B.J.; Casellas, P.; Blythman, H.E.; Jansen, F.K. Study of the plasma clearance of antibody—Ricin-A-chain immunotoxins. Evidence for specific recognition sites on the A chain that mediate rapid clearance of the immunotoxin. *Eur. J. Biochem.* **1986**, *155*, 1–10. [CrossRef]

273. Fulton, R.J.; Uhr, J.W.; Vitetta, E. In Vivo Therapy of the BCL1 Tumor: Effect of Immunotoxin Valency and Deglycosylation of the Ricin A Chain. *Cancer Res.* **1988**, *48*, 2626–2631.

274. Blakey, D.C.; Watson, G.; Knowles, P.P.; Thorpe, P. Effect of chemical deglycosylation of ricin A chain on the in vivo fate and cytotoxic activity of an immunotoxin composed of ricin A chain and anti-Thy 1.1 antibody. *Cancer Res.* **1987**, *47*, 947–952.

275. Street, N.E.; Fulton, R.J.; Sanders, V.M.; Vitetta, E.S. Inhibition of the helper function of murine T cells with Fab'-anti-L3T4 ricin A chain immunotoxin. *J. Immunol.* **1987**, *139*, 1734–1738.

276. Li, C.; Yan, R.; Yang, Z.; Wang, H.; Zhang, R.; Chen, H.; Wang, J. BCMab1-Ra, a novel immunotoxin that BCMab1 antibody coupled to Ricin A chain, can eliminate bladder tumor. *Oncotarget* **2017**, *8*, 46704–46705. [CrossRef]

277. Petros, R.A.; DeSimone, J.M. Strategies in the design of nanoparticles for therapeutic applications. *Nat. Rev. Drug Discov.* **2010**, *9*, 615–627. [CrossRef] [PubMed]

278. Parveen, S.; Misra, R.; Sahoo, S.K. Nanoparticles: A boon to drug delivery, therapeutics, diagnostics and imaging. *Nanomedicine* **2012**, *8*, 147–166. [CrossRef] [PubMed]

279. Nichols, J.W.; Bae, Y.H. Odyssey of a cancer nanoparticle: From injection site to site of action. *Nano Today* **2012**, *7*, 606–618. [CrossRef]

280. Skotland, T.; Iversen, T.-G.; Sandvig, K. Development of nanoparticles for clinical use. *Nanomedicine* **2014**, *9*, 1295–1299. [CrossRef] [PubMed]

281. Tekle, C.; van Deurs, B.; Sandvig, K.; Iversen, T.-G. Cellular trafficking of quantum dot-ligand bioconjugates and their induction of changes in normal routing of unconjugated ligands. *Nano Lett.* **2008**, *8*, 1858–1865. [CrossRef]

282. Iversen, T.-G.; Frerker, N.; Sandvig, K. Quantum dot bioconjugates: Uptake into cells and induction of changes in normal cellular transport. In *Colloidal Quantum Dots for Biomedical Applications IV*; International Society for Optics and Photonics: Bellingham, WA, USA, 2009; Volume 7189, p. 71890T.

283. Li, Y.; Liu, W.; Sun, C.; Zheng, M.; Zhang, J.; Liu, B.; Wang, Y.; Xie, Z.; Xu, N. Hybrids of carbon dots with subunit B of ricin toxin for enhanced immunomodulatory activity. *J. Colloid Interface Sci.* **2018**, *523*, 226–233. [CrossRef] [PubMed]

284. Díaz, R.; Pallarès, V.; Cano-Garrido, O.; Serna, N.; Sánchez-García, L.; Falgàs, A.; Pesarrodona, M.; Unzueta, U.; Sánchez-Chardi, A.; Sánchez, J.M.; et al. Selective CXCR4$^+$ Cancer Cell Targeting and Potent Antineoplastic Effect by a Nanostructured Version of Recombinant Ricin. *Small* **2018**, *14*, 1800665. [CrossRef]

285. Audi, J.; Belson, M.; Patel, M.; Schier, J.; Osterloh, J. Ricin poisoning: A comprehensive review. *JAMA* **2005**, *294*, 2342–2351. [CrossRef]

286. Smallshaw, J.E.; Vitetta, E.S. Ricin Vaccine Development. In *Ricin and Shiga Toxins: Pathogenesis, Immunity, Vaccines and Therapeutics*; Mantis, N., Ed.; Current Topics in Microbiology and Immunology; Springer: Berlin/Heidelberg, Germany, 2012; pp. 259–272, ISBN 978-3-642-27470-1.

287. Porter, A.; Phillips, G.; Smith, L.; Erwin-Cohen, R.; Tammariello, R.; Hale, M.; DaSilva, L. Evaluation of a ricin vaccine candidate (RVEc) for human toxicity using an in vitro vascular leak assay. *Toxicon* **2011**, *58*, 68–75. [CrossRef]

288. Bascon, J.U. Vascular leak syndrome: A troublesome side effect of immunotherapy, Immunopharmacology, 39/3 (1998) 255. *Immunopharmacology* **1998**, *39*, 255–257.

289. Vance, D.J.; Rong, Y.; Brey, R.N.; Mantis, N.J. Combination of Two Candidate Subunit Vaccine Antigens Elicits Protective Immunity to Ricin and Anthrax Toxin in Mice. *Vaccine* **2015**, *33*, 417–421. [CrossRef] [PubMed]

290. Vance, D.J.; Mantis, N.J. Progress and Challenges Associated with the Development of Ricin Toxin Subunit Vaccines. *Expert Rev. Vaccines* **2016**, *15*, 1213–1222. [CrossRef]

291. Wahome, N.; Sully, E.; Singer, C.; Thomas, J.C.; Hu, L.; Joshi, S.B.; Volkin, D.B.; Fang, J.; Karanicolas, J.; Jacobs, D.J.; et al. Novel Ricin Subunit Antigens with Enhanced Capacity to Elicit Toxin-Neutralizing Antibody Responses in Mice. *J. Pharm. Sci.* **2016**, *105*, 1603–1613. [CrossRef]

292. Sully, E.K.; Whaley, K.J.; Bohorova, N.; Bohorov, O.; Goodman, C.; Kim, D.H.; Pauly, M.H.; Velasco, J.; Hiatt, E.; Morton, J.; et al. Chimeric plantibody passively protects mice against aerosolized ricin challenge. *Clin. Vaccine Immunol.* **2014**, *21*, 777–782. [CrossRef] [PubMed]

293. Yermakova, A.; Klokk, T.I.; Cole, R.; Sandvig, K.; Mantis, N.J. Antibody-Mediated Inhibition of Ricin Toxin Retrograde Transport. *mBio* **2014**, *5*, e00995-13. [CrossRef] [PubMed]

294. Herrera, C.; Vance, D.J.; Eisele, L.E.; Shoemaker, C.B.; Mantis, N.J. Differential neutralizing activities of a single domain camelid antibody (VHH) specific for ricin toxin's binding subunit (RTB). *PLoS ONE* **2014**, *9*, e99788. [CrossRef]

295. Vance, D.J.; Tremblay, J.M.; Rong, Y.; Angalakurthi, S.K.; Volkin, D.B.; Middaugh, C.R.; Weis, D.D.; Shoemaker, C.B.; Mantis, N.J. High-Resolution Epitope Positioning of a Large Collection of Neutralizing and Nonneutralizing Single-Domain Antibodies on the Enzymatic and Binding Subunits of Ricin Toxin. *Clin. Vaccine Immunol.* **2017**, *24*. [CrossRef]

296. Poon, A.Y.; Vance, D.J.; Rong, Y.; Ehrbar, D.; Mantis, N.J. A Supercluster of Neutralizing Epitopes at the Interface of Ricin's Enzymatic (RTA) and Binding (RTB) Subunits. *Toxins* **2017**, *9*, 378. [CrossRef]

297. O'Hara, J.M.; Mantis, N.J. Neutralizing Monoclonal Antibodies against Ricin's Enzymatic Subunit Interfere with Protein Disulfide Isomerase-Mediated Reduction of Ricin Holotoxin In Vitro. *J. Immunol. Methods* **2013**, *395*, 71–78. [CrossRef]

298. Pincus, S.; Das, A.; Song, K.; Maresh, G.; Corti, M.; Berry, J. Role of Fc in Antibody-Mediated Protection from Ricin Toxin. *Toxins* **2014**, *6*, 1512–1525. [CrossRef]

299. Yermakova, A.; Mantis, N.J. Protective Immunity to Ricin Toxin Conferred by Antibodies against the Toxin's Binding Subunit (RTB). *Vaccine* **2011**, *29*, 7925–7935. [CrossRef] [PubMed]

300. Yermakova, A.; Mantis, N.J. Neutralizing activity and protective immunity to ricin toxin conferred by B subunit (RTB)-specific Fab fragments. *Toxicon* **2013**, *72*, 29–34. [CrossRef] [PubMed]

301. Yermakova, A.; Vance, D.J.; Mantis, N.J. Sub-domains of ricin's B subunit as targets of toxin neutralizing and non-neutralizing monoclonal antibodies. *PLoS ONE* **2012**, *7*, e44317. [CrossRef]

302. Cherubin, P.; Quiñones, B.; Teter, K. Cellular recovery from exposure to sub-optimal concentrations of AB toxins that inhibit protein synthesis. *Sci. Rep.* **2018**, *8*, 2494. [CrossRef] [PubMed]

303. Lin, J.-Y.; Tserng, K.-Y.; Chen, C.-C.; Lin, L.-T.; Tung, T.-C. Abrin and Ricin: New Anti-tumour Substances. *Nature* **1970**, *227*, 292. [CrossRef] [PubMed]

304. Lin, J.Y.; Chang, Y.C.; Huang, L.Y.; Tung, T.C. The cytotoxic effects of abrin and ricin on Ehrlich ascites tumor cells. *Toxicon* **1973**, *11*, 379–381. [CrossRef]

305. Fodstad, O.; Kvalheim, G.; Godal, A.; Lotsberg, J.; Aamdal, S.; Høst, H.; Pihl, A. Phase I study of the plant protein ricin. *Cancer Res.* **1984**, *44*, 862–865.

306. Davies, M.P.A.; Barraclough, D.L.; Stewart, C.; Joyce, K.A.; Eccles, R.M.; Barraclough, R.; Rudland, P.S.; Sibson, D.R. Expression and splicing of the unfolded protein response gene XBP-1 are significantly associated with clinical outcome of endocrine-treated breast cancer. *Int. J. Cancer* **2008**, *123*, 85–88. [CrossRef]

307. Koong, A.C.; Chauhan, V.; Romero-Ramirez, L. Targeting XBP-1 as a novel anti-cancer strategy. *Cancer Biol. Ther.* **2006**, *5*, 756–759. [CrossRef]

Review

How Ricin Damages the Ribosome

Przemysław Grela *, Monika Szajwaj, Patrycja Horbowicz-Drożdżal and Marek Tchórzewski

Department of Molecular Biology, Maria Curie-Skłodowska University, Akademicka 19, 20-033 Lublin, Poland;
monika.szajwaj@poczta.umcs.lublin.pl (M.S.); patrycja.horbowicz@gmail.com (P.H.-D.);
maro@hektor.umcs.lublin.pl (M.T.)
* Correspondence: przemek@hektor.umcs.lublin.pl; Tel.: +48-81-537-5954

Received: 31 March 2019; Accepted: 24 April 2019; Published: 27 April 2019

Abstract: Ricin belongs to the group of ribosome-inactivating proteins (RIPs), i.e., toxins that have evolved to provide particular species with an advantage over other competitors in nature. Ricin possesses RNA N-glycosidase activity enabling the toxin to eliminate a single adenine base from the sarcin-ricin RNA loop (SRL), which is a highly conserved structure present on the large ribosomal subunit in all species from the three domains of life. The SRL belongs to the GTPase associated center (GAC), i.e., a ribosomal element involved in conferring unidirectional trajectory for the translational apparatus at the expense of GTP hydrolysis by translational GTPases (trGTPases). The SRL represents a critical element in the GAC, being the main triggering factor of GTP hydrolysis by trGTPases. Enzymatic removal of a single adenine base at the tip of SRL by ricin blocks GTP hydrolysis and, at the same time, impedes functioning of the translational machinery. Here, we discuss the consequences of SRL depurination by ricin for ribosomal performance, with emphasis on the mechanistic model overview of the SRL *modus operandi*.

Keywords: ribosome-inactivating protein; ricin; ribosome; sarcin-ricin loop (SRL); GTPase associated center (GAC); P-proteins; translation

Key Contribution: This review summarizes the current knowledge of ricin-induced impairment of the ribosome and translation process and highlights important aspects in this field.

---

1. Introduction

Toxic proteins are naturally present in a wide variety of species [1]. It is thought that they have evolved to play a specific role in defense against animals, pathogens, and various insects, giving advantage in a particular niche [2,3]. One of the most toxic proteins known in nature are plant toxins from the class of ribosome-inactivating proteins. The mechanism of toxicity of plant toxins is of great interest because they are present in human foods [4,5] and used in ethnomedicine [6] and cosmetics [7]. Currently, they have attracted attention in broad biotechnological applications [8,9]. Importantly, they also pose a significant threat, as numerous plant toxins, such as ricin, have been recognized as a biological weapon [10] and are recently considered as an agent of bioterrorism [11]. Ricin is one of the most common and potent lethal biological molecules known [9]. Ricin and related proteins display one common feature, i.e. the ability to inhibit ribosomes, and this particular feature laid the foundation for the general name for all these proteins: "ribosome-inactivating proteins" (RIPs) [12]. In general, all RIPs have been classified as single chain (type 1) and double chain (type 2) proteins [13]. Type 1 RIPs consist of an enzymatically active chain (A) and include for instance: pokeweed antiviral protein (PAP), trichosanthin, saporin, gelonin, and luffin. Ricin, abrin, and Shiga toxin are classified as type 2 RIPs, where the enzymatically active A-chain is disulfide-linked to a B-chain, which provides an active chain with higher ability to enter cells. Thus, type 2 RIPs are considerably more toxic than type 1 RIPs [13]. Also, a third class of RIPs, termed type 3, has been recognized. It includes only a few members, with the prominent example of jasmonate induced protein (JIP60) [14] and maize b-32 protein, which requires proteolytic process to become active [15]. In 2010, a new nomenclature has been proposed for RIPs, in which they are termed A (type 1), AB (type 2), AC (type 3, similar to JIP60), and AD (Type 3, similar to b32) [16]. The current model of the catalytic activity of ricin and other RIPs assumes that, using the N-glycosidase activity, the toxin removes a single adenine base from the ribosomal RNA (rRNA) located at the heart of the ribosomal GTPase associated center (GAC) recognized as a central ribosomal element responsible for fueling of the translational machinery [17]. The prime target of ricin is a structural element of rRNA and is called the sarcin–ricin loop (SRL) because this element is targeted by another ribo-toxin, α-sarcin; yet in this case, the SRL undergoes endonucleolytic cleavage [18]. The ricin-dependent depurination of the SRL exerts a deleterious effect, blocking the ribosome action and thereby hampering protein synthesis. It should be stressed that despite the high homology of the SRL within the three domains of life, mainly eukaryotic ribosomes undergo efficient depurination, however, several type I RIPs were shown to depurinate prokaryotic ribosomes as well [19]. The extraordinary specificity of the RIP action is based on their interaction with unique eukaryotic proteins; in the case of ricin, the ribosomal P-proteins are regarded as a main docking part [20–22] while the L3 protein represents the main landing platform for the PAP [23,24]. However, regardless of the docking element, the eukaryotic ribosomal proteins are considered as guide molecules for RIPs, directing and activating the toxin toward its catalytic target - the SRL. The toxicity of ricin was mainly associated with inhibition of protein synthesis, but it should be stressed that there is no clear link between depurination of SRL, ribosome hampering and the toxic effect in the cell. The currently available information concerning ricin toxicity suggests that the toxic effect on cell metabolism is multifactorial and involves induction of numerous pathways that lead to cell death; however, molecular details elucidating the effect of ricin on cell metabolism are still elusive [25].

The numerous biochemical approaches [26–29] and especially current structural analyses have provided deep insight into the role of SRL in ribosome performance and cast light on the molecular consequences of its depurination-induced damage. Here, we are presenting the current understanding of the structure and function of SRL during the translational cycle, with particular focus on the molecular consequences of ricin-dependent adenine base removal on ribosome performance.

2. Ribosome Structure as A Prime Target for Ricin

2.1. Ribosome and Its Active Sites

Protein biosynthesis, being a critical biological pathway, provides a suitable target for many toxins or natural inhibitors with the ribosome as the main objective [30,31]. The ribosome is one of the most conserved and sophisticated macromolecular machines of the cell in all domains of life [32]. It is composed of a small and a large subunit, which together form a fully functional ribosome. Ribosomal subunits are composed of ribosomal RNA (rRNA) and a large number of ribosomal proteins, but it is the rRNA that plays the most critical functional role, defining the ribosome as a ribozyme [33]. The ribosome can be regarded as a scaffolding platform for many molecules, e.g., mRNA, aminoacyl-tRNAs (aa-tRNA), and protein factors, which together form the translational machinery converting genetic information into functional proteins [34]. The ribosome, being the central element of this machinery, carries out its task through several functional elements. The small subunit decodes the genetic information delivered by mRNA, whereas the large subunit hosts the catalytic peptidyl transferase center (PTC), where amino acids delivered by aa-tRNAs are linked into polypeptides [32]. Additionally, three tRNA binding sites can be recognized on the ribosome: aminoacyl (A), peptidyl (P), and exit (E) sites. The A site accommodates the incoming aa-tRNA, the P site binds the peptidyl-tRNA thus carrying the nascent polypeptide chain, and the E site binds deacylated tRNA before it dissociates from the ribosome. On the large ribosomal subunit there is also the GTPase associated center (GAC) which belongs to the ribosomal elements responsible for administrating the continuous motions of the ribosome along the translational cycle.

2.1.1. The GTPase Associated Center

The translation process to proceed efficiently and to comply with the metabolic needs of the cell requires many protein factors, which sequentially guide the ribosome through the protein synthesis cycle. The most critical group of factors are proteins that bind and hydrolyze GTP, called translational GTPases (trGTPases), which confer the unidirectional trajectory of the translational machinery at the expense of energy released from GTP hydrolysis [17]. The landing platform for all trGTPases is situated on the large ribosomal subunit and is called the GTPase associated center (GAC) (Figure 1) [35].

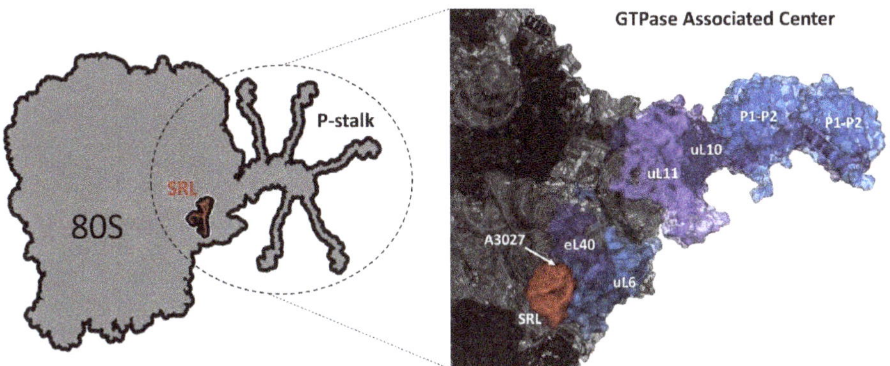

Figure 1. Model of GTPase associated center (GAC). Left panel: scheme of the 80S ribosome with the ribosomal P-stalk shown with extended C-terminal regions and the sarcin–ricin loop (SRL) (red). Right panel: fragment of 60S *S. cerevisiae* ribosome 25S rRNA (PDB code 3U5H) and 60S subunit (PDB code 3U5I); 25S rRNA and ribosomal proteins are indicated as light gray and dark gray colors, respectively. The fitted schematic structure of uL10 protein fragment in complex with the N-terminal domains of P1/P2-proteins (PDB code 3A1Y) from Archaea is depicted as dark blue and marine blue, respectively. The position of the yeast 60S subunit is oriented to show A_{3027} of the SRL and positions of uL11 (slate blue), uL40 (deep blue) and uL6 (sky blue) proteins. Model prepared with PyMol software (The PyMOL Molecular Graphics System, version 1.5.0.4, Schrödinger, LLC, NY, USA).

The GAC is responsible for recruitment of trGTPases and stimulation of factor-dependent GTP hydrolysis [36]. The center consists of two main functional elements: a conserved fragment of rRNA called the sarcin–ricin loop (SRL) and the ribosomal stalk, composed exclusively of ribosomal proteins [37]. Both elements are critical for activation of trGTPases [28,38], and mutual cooperativity of the SRL and the stalk elements has been shown to be pivotal in stimulation of GTP hydrolysis by trGTPases [39–41].

The Sarcin–Ricin Loop

The sarcin–ricin loop is one of the most conserved rRNA regions of the ribosome, which underlines its importance in ribosome function. It is located in helix 95, in domain VI of 23S/25S/28S rRNA (nucleotides 2646-2674 in *E. coli*, 3012-3042 in yeast) (Figure 2A).

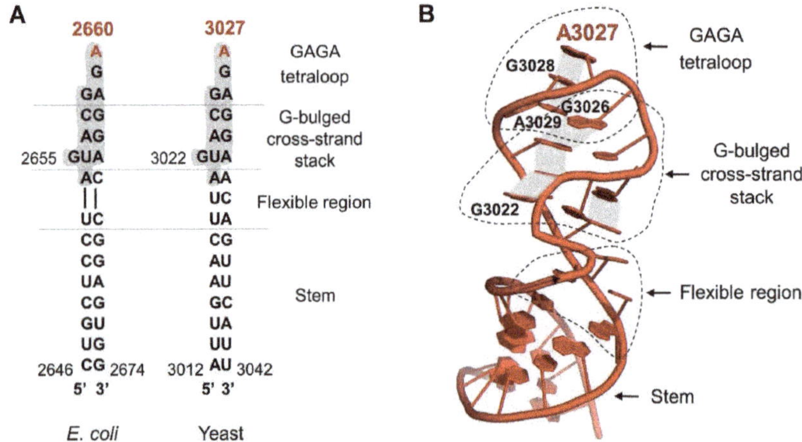

Figure 2. The model of sarcin–ricin loop structure. (**A**) Alignment of highly conserved secondary structures of yeast and *E. coli* SRL. The red color indicates the key adenine hydrolyzed by the ricin. A conserved fragment of 12 nucleotides is marked with a gray color. (**B**) Structure of the *S. cerevisiae* SRL (PDB code 3U5H). The key adenine was marked in red (A_{3027}). The individual structural elements of the SRL - the stem, the flexible region, the G-bulged cross-strand stack with the highlighted individual nucleotides and the GAGA loop - are marked with a dotted black line. The gray fields show the non-canonical π-stacking interactions between particular bases. Model prepared by PyMol software, (The PyMOL Molecular Graphics System, version 1.5.0.4, Schrödinger, LLC, NY, USA) based on [42].

Unlike other rRNA regions, which form compact structures coordinated by rRNA–rRNA or rRNA-protein interactions, the SRL exists on the ribosome as an autonomous unit [43–45] and is exposed to the solvent [46]. This unique feature is critical for its accessibility for external factors like trGTPases. From the structural point of view, the SRL has the conformation of a distorted hairpin [47]. It consists of several well-organized elements (Figure 2B) [48]: the stem, the flexible region, the G-bulged cross-strand stack, and the GAGA loop [45] with the key adenine base (A_{2660}/A_{3027}—*E. coli*/*S.cerevisiae* numbering), which is a target for ricin activity. The SRL stem structure is formed mainly by classical Watson–Crick base pairs, whereas the rest of the structure is stabilized mainly by π-stacking interactions. The critical element, the GAGA tetra-loop, forms a compact well-organized structure [43]. The spatial organization of the loop structure is determined by non-canonical interactions between base pairs, allowing the carbohydrate-phosphate backbone of RNA to form the atypical spatial form, which is recognized by translational factors [43]. The critical bases A_{2660}, G_{2661}, and A_{2662} in the GAGA loop (*E. coli* numbering) associate with each other via non-canonical π-stacking interactions that stabilize the loop structure [42,43]. The crucial A_{2660}, located at the top of the hairpin structure, is stabilized via the stacking interactions with G_{2661} (Figure 2B). Importantly, A_{2660} is fully exposed and does not form any hydrogen bonds with other bases of the loop, making it easily accessible to external factors like ricin [49]. Biochemical studies on prokaryotic and eukaryotic ribosomes have shown that the SRL represents a critical element responsible for the interaction and stimulation of all trGTPases activity [50,51]. Additionally, it has been early recognized that this structure represents the main target for numerous toxins [50,52–54]. Recent structural analyses have brought detailed insight into the intricate interplay between the SRL and trGTPases and at the same time cast light on the molecular aspects of ricin toxicity [28,29,40,55]. In general, all trGTPases interact with the SRL via the GTP-binding domain (G-domain) comprising the active site of the factor, responsible for GTP hydrolysis [17]. It should be stressed that the structure of the G-domain is evolutionarily conserved among all trGTPases, and biochemical and structural investigations support the notion that the mechanism of activation of GTP hydrolysis by the ribosome is universally conserved [17,56]. The trGTPases convert chemical energy into mechanical forces at the expense of GTP hydrolysis, and this drives the ribosome through the translational cycle. The key role in catalysis is played by an invariant histidine residue (His_{84} in EF-Tu, His_{87} in EF-G, or His_{61} in SelB, or His_{108} in eEF2) [56–60]. This histidine may adopt two conformations: an inactive "flipped-out" state (pointing away from the γ-phosphate of GTP) and an active flipped-in state (reaching towards the γ-phosphate) (Figure 3). After joining the factor to the ribosome, the SRL is "inserted" into the catalytic center of the G-domain, and the phosphate moiety of A_{2662} coordinates, by means of electrostatic interactions, the catalytic histidine positioning to the active "flipped-in" position towards the γ-phosphate of GTP [28,38,61]. The positively charged histidine points towards the water molecule aimed at the nucleophilic attack on GTP γ-phosphate [29,38,58,62]. Additionally, the phosphate of A_{2662} coordinates a Mg^{2+} ion, important in positioning of the Asp residue (Asp_{21} in EF-Tu, Asp_{22} in EF-G, Asp_{10} in SelB), which is also crucial for GTP hydrolysis (Figure 3) [28,58].

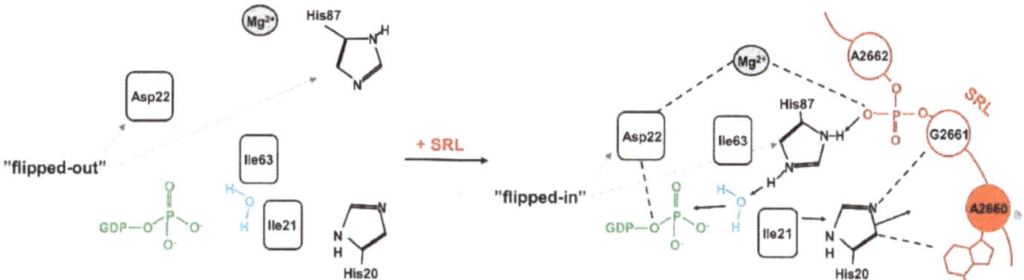

Figure 3. Model of GTP hydrolysis activation with the aid of sarcin–ricin loop (SRL), with the EF-G as trGTPase. Left panel: the organization of the active site in isolated EF-G. Asp_{22} and His_{87} are in "flipped-out" state, pointing away from GTP; the "hydrophobic gate" formed by amino acids Ile_{63} and Ile_{21} prevents His_{87} from adopting the active conformation. Right panel: the reorganization of the active site of EF-G as a result of binding of EF-G to ribosome and inserting the SRL to the G-domain. Base A_{2660} interacts with His_{20}, which induces Ile_{21} movement away from GTP and "hydrophobic gate" opening. The phosphate of A_{2662} directs His_{87} and Asp_{22} residues (through Mg^{2+}) to "flipped-in" conformation which allows for water molecule activation and GTP hydrolysis (see the details in the text).

Recently, A_{2662} and G_{2661} within the tetraloop structure were distinguished as critical elements, directly involved in stimulation of the GTP hydrolysis process [28,38]. Interestingly, α-sarcin cleaves the bond between eukaryotic nucleotide equivalents to *E. coli* G_{2661} and A_{2662}, which results in ribosome inactivation [63,64]. The A_{2660} base, which is cleaved-off by ricin, plays a distinct but weighty function, which could be named as the "power behind the throne" title role. As shown based on the structure of the EF-G–ribosome complex in a pre-translocation state, an intricate network of hydrogen-based interactions involving the G_{2661} and A_{2660} of the SRL, EF-G (Glu_{456}, Arg_{660}, Ser_{661} and Gln_{664}), and ribosomal L6 (Lys_{175}) is formed in the immediate vicinity of the GTPase active site, with A_{2660} being the central element of the network [40]. Thus, depurination of A_{2660} may prevent the surrounding elements from adopting the active conformation, which is required to bind the metal ions necessary to stabilize Asp_{22} and neighboring regions of EF-G in the activated form [55,65]. On the other hand, A_{2660}, together with G_{2661}, plays a crucial role in opening of a so-called hydrophobic gate, which prevents the invariant His residue in the free trGTPase from achieving an active state and spontaneous GTP hydrolysis. In the complex of the EF-G/ribosome, the bases A_{2660} and G_{2661} interact with His_{20} of EF-G, which in turn interacts with Ile_{21}, forming a hydrophobic gate with Ile_{63}. These interactions contribute to its opening and repositioning the His_{87} into its active position [40]. The structural analyses of the A_{2660} role are supported by biochemical insight. It has been shown that lack of the single exocyclic N6 amino group at position 2660 within rRNA inhibited GTP hydrolysis on the EF-G/ribosome complex. Importantly, the introduction of different exocyclic groups with dissimilar chemical groups, such as inosine, dimethyladenosine, or even 6-methylpurine, restored the GTP hydrolysis activity of EF-G. The experimental biochemical data indicate that the critical favorable chemical feature is related to electron configuration that allows participation in the aromatic π-electron interaction system of the purine, which in turn facilitates the π-stacking effect [66]. It should be underlined that despite the vast number of data collected from only the bacterial model, the amino acid sequence, called PGH motif (with the invariant His residue) is universally conserved and present in EF-Tu, EF-G, IF2, and RF3, as well as archeal and eukaryotic trGTPases [67]. What is more, the highly conserved A_{2660} base moiety (A_{3027} - yeast, A_{4324} - rat, A_{4605} - human) within the tetra-loop of SRL [56] represents the most crucial base which contributes to a cooperative interaction network, which stabilizes the active state of trGTPases, promoting GTP hydrolysis.

The Ribosomal Stalk

The ribosomal stalk represents a vital element within the ribosomal GTPase associated center. Stalk structure is composed of two distinct parts - the base of the stalk and its lateral elements [68]. The stalk base is constituted by conserved ribosomal proteins uL11 and uL10, which anchor the stalk to the rRNA [69,70]. The lateral part of the stalk has multimeric architecture and is built of multi-dimeric protein elements, which are unique for bacteria and eukaryotes. In bacteria, the bL12 proteins form a dimer, which is regarded as a basic structural element, and two, three, and even four dimers can form the lateral part anchored to the ribosome through uL10 [35,71]. In eukaryotes, the P1 and P2 proteins form a dimer, and two dimers are linked to the uL10 protein, forming pentameric architecture called the P-stalk, uL10-(P1-P2)$_2$ [37,69,70,72–75]. It should be underlined that the stalk fulfils the same function on the ribosome, irrespectively of the life-domain origin, namely participation in stimulation of GTP hydrolysis [68]; however, the bL12 and P1/P2 proteins are not evolutionarily related and are regarded as analogous proteins [76,77]. The stalk is the only structure on the ribosome composed of multiple proteins. The eukaryotic stalk architecture has a complex nature. It is constituted by two P1/P2 protein dimers; each dimer is built of two domains: an N-terminal globular domain (NTD), responsible for dimerization and anchoring the dimer to uL10, and an unstructured C-terminal domain (CTD), regarded as a functional part interacting with trGTPases [70,73,78]. Both elements are connected through a highly flexible hinge region [73,79]. The most prominent feature of the eukaryotic P1/P2 stalk proteins is the highly conserved element present at the CTD, composed of a stretch of acidic and hydrophobic amino acids (EESEESDDDMGFGLFD) and regarded as the main functional element of the stalk. This element is involved in the interaction with trGTPases and toxins such as ribosome-inactivating proteins (RIPs) [22,80–84]. A unique feature of the eukaryotic stalk is multiplication of CTDs. The conserved CTD is also found on the uL10; therefore, five CTDs are present on the stalk: four coming from two P1/P2 dimers and one from uL10. The phenomenon of CTDs multiplication was functionally coupled with the qualitative aspect of ribosome action related to maintenance of translation accuracy [85]. It was proposed that the multiple CTDs might accelerate interaction with eEF1A, which is regarded as trGTPase with the highest GTP hydrolysis turnover. Interestingly, this feature has been hijacked by RIP toxins, and it has been shown that multiplication of P1/P2 proteins increase the interaction rate of the toxin [86].

2.2. Mode of Ricin Interaction with Ribosome

It has been established that ricin inhibits translation through its ability to remove/depurinate a specific adenine base of the universally conserved SRL [87], which is a crucial part of the GAC on the ribosome [27,38,88,89]. The SRL has been found as a primary target for ricin and other RIPs, and the specificity of the interaction with eukaryotic ribosomal proteins plays a critical role in ricin catalytic activity towards SRL. As shown over two decades ago, the efficiency of rRNA depurination in the intact ribosome is much greater than the depurination of isolated 28S rRNA. The k_{cat} of ricin against naked rRNA is more than 4 orders of magnitude lower than that of rRNA constituting a part of the ribosome [90–92]. Ricin depurinates the naked 23S rRNA from *E. coli* SRL, but not the intact ribosomes from *E. coli* [91], showing at the same time extraordinary specificity towards intact eukaryotic ribosomes [20,21,93], what underlines the role of ribosomal proteins in the process. The same applies to other related RIPs, such as Shiga toxin 1 (Stx1) [93,94], Shiga toxin 2 (Stx2) [95,96], trichosanthin (TCS) [97–99], and maize RIP [100], which specifically depurinate the SRL on the eukaryotic ribosome. In the case of ricin, the mechanistic model of molecular recognition of the ribosome assumes a double-step mechanism, involving first slow and nonspecific electrostatic-based interactions with the ribosome and then fast specific interactions based on the ribosomal stalk interplay, leading to its attack on the SRL rRNA [90]. Although the SRL is highly conserved among ribosomes in all species, the P-proteins determine the specificity of ricin and other RIPs toward eukaryotic ribosomes [12,22]. The deletion of stalk P-proteins from ribosomes greatly reduces the depurination activity and cellular sensitivity to ricin, indicating that binding to the P-stalk is a critical step in depurination of the SRL and in the toxicity of ricin [101–103]. The structural investigations provide significant insights into the mode of interaction between P-proteins and the ricin or trichosanthin (TCS), which hijacks the translational factor recruitment function of the ribosomal P-stalk to reach its target site on the ribosome [22,84]. Especially, the interaction site of P-proteins with RIPs was mapped to a short conserved 11-mer peptide, SDDDMGFGLFD, present at the CTDs of all P-proteins [98]. This interaction is required for the full activity of ricin and other RIPs, and biochemical analyses confirmed that positively charged residues, especially the cluster of arginines, play a key role [101]. As was shown in a TCS study, the interaction of TCS is primarily mediated by the electrostatic interactions of K_{173}, R_{174}, and K_{177} in the C-terminal domain of TCS with the conserved DDD residues in the CTDs of P-proteins. However, hydrophobic interactions also play a vital role in stabilization of the bilateral interplay between TCS and the conserved C-terminal peptide of P-proteins [93,99]. The tertiary structure of the catalytic subunit of ricin (RTA) with a short peptide corresponding to the last six conserved residues of the stalk proteins (GFGLFD) showed that the peptide docks into a hydrophobic pocket at the C-terminus of RTA [84,104]. The structural superposition of TCS-P-protein and RTA-P-protein complexes demonstrated that the short C-terminal peptide of P-proteins adopts distinct orientations and slightly different interaction modes with the two different RIPs, suggesting that the flexibility of the CTD facilitates accommodation of different class of RIPs to the ribosome [84,104]. The kinetic studies showed that the P1-P2 heterodimeric conformation of P-proteins in the stalk pentamer represents an optimal binding site for RTA, where individual P-protein CTDs play non-equivalent roles with a pivotal role of P1 CTD [65]. Additionally, previous results obtained using yeast as an experimental system showed that the two dimers, P1A-P2B and P1B-P2A, do not interact equally with RTA [102], suggesting that these dimers may have a different architecture and their CTDs may not be equally accessible to external factors, such as RTA or other RIPs. The high specificity of ricin interaction with the P-stalk is also reflected by the measured dissociation constant, which is in a nanomolar range [65,80,86]. Thus, the kinetic model of RTA interaction with the ribosome and SRL depurination assumes that the toxin initially interacts with the P-protein stalk, and it allows orienting the active site of the toxin toward the SRL, which in turn places it in correct orientation for binding to the target adenine. It is also proposed that the P-stalk binding event allosterically stimulates the catalysis of ribosome depurination by RTA, explaining the extraordinary specificity of the toxin toward eukaryotic ribosomes [101].

3. Toxic Action of Ricin on The Translational Process

Ricin is composed of two subunits, RTA and RTB, covalently linked through a disulfide bond. In the form of holotoxin, it does not exhibit catalytic activity toward ribosome [105]. When RTA is separated from RTB, the cluster of arginine residues located at the interface domain between RTA and RTB is exposed to the solvent and serves as an interaction platform for P-stalk proteins. The RTA-stalk interaction stimulates the toxin to trigger its enzymatic activity by orienting the active site of RTA (opposite to the arginine interface) toward the SRL [101]. RTA is an RNA N-glycosidase (EC 3.2.2.22) that hydrolyzes the N-glycosidic bond between a specific adenine on the SRL and the sugar backbone [91]. The specificity of rRNA depurination by RTA is determined by the conformation of the topical part of the SRL loop structure, and the GAGA sequence with the prominent key adenine base is recognized as the major element [106,107]. During the catalysis of the SRL depurination process by ricin, the conserved adenine on the tip of the sarcin–ricin loop is inserted between two Tyr residues (in the RTA catalytic center - Tyr_{80} and Tyr_{123}) to form π-stacking interactions [108,109]. Additionally, the adenine position is stabilized by hydrogen binding with RTA Gly_{121}, Val_{81}, Glu_{177}, and Arg_{180} residues [110]. It has been shown that two RTA residues, i.e. Glu_{177} and Arg_{180}, play a crucial role in the hydrolysis of N-glycosidic bonds by stabilizing the transition state during catalysis of the depurination reaction (Figure 4) [111,112].

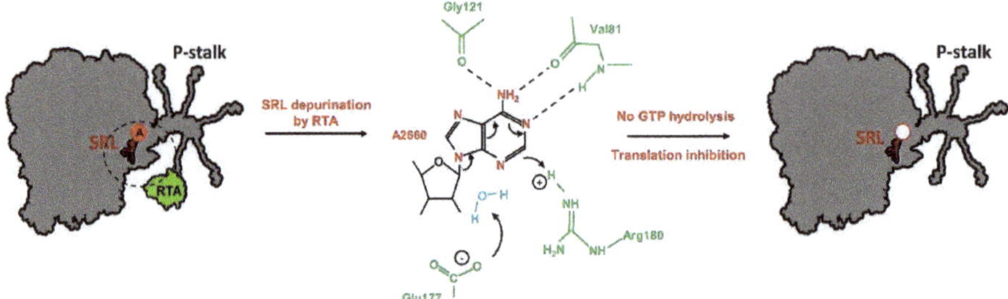

Figure 4. Model of sarcin–ricin loop (SRL) depurination by RTA. Removal of key A_{2660} residue (*E. coli* numbering) impairs the intricate interaction network responsible for stabilization of the active state of trGTPases, resulting in the inhibition of translation process. The base of A_{2660} is bound to the catalytic center of RTA with π-stacking interactions and its position is stabilized by hydrogen bonds with Gly_{121}, Val_{81} and Arg_{180} (green). Glu_{177} and Arg_{180} (green) play crucial role in catalysis by transition state stabilization. Arg_{180} residue protonates base of A_{2660} causing delocalization of ring electrons. Glu_{177} polarizes water molecule (blue) and resultant hydroxide ion attacks positive center on ribose which leads to hydrolysis of N-glycosidic bond; prepared based on [110].

As already discussed, the structural stability of the SRL is provided mostly by the π-stacking interaction network [52,113,114]. Removal of the key adenine at the tip of the SRL may destabilize this type of interactions, thereby affecting the SRL structure stability and abolishing an extended interaction network responsible for stabilization of the active state of trGTPases (Figure 4).

It was already observed in the 1970s that ricin inhibits translation in mammalian cells [115], as confirmed in an in vitro experimental system [53,116–119]. Early analyses in in vitro protein synthesis systems have shown that the presence of ricin blocks the synthesis of polypeptides, and this was mainly associated with the elongation step of the translational cycle [116,117,120–122]. It has also been reported that ricin does not affect the synthesis of peptide bonds [53,116,123,124], but a significant inhibition of GTP hydrolysis was associated with ricin action [53,54,116,124–130]. Numerous analyses have shown that eEF2, i.e. a factor involved in the translocation event during the elongation cycle, binds less efficiently to ribosomes modified by ricin [54,125,126,128,131]. Additionally, it has also been shown that treatment of ribosomes with ricin decreased the level of GTP hydrolysis by the eEF2 factor [53,116]. Further analyses demonstrated that ricin inhibited translocation, and the effect was dependent on the eEF2 concentration used [123]. Additional evidence for the inhibitory effects of ricin on translocation was provided by applying test with the use of diphtheria toxin, showing that the ribosomes treated with both toxins mainly paused at the beginning of the mRNA [119]. It was also shown that the rate of formation of pre-initiation complexes, i.e. the attachment of the 60S subunit to the pre-initiation complex 40S, was decreased under the influence of ricin [119]. Thus, these analyses have provided evidence that the modification of the 60S subunit by ricin resulted not only in inhibition of elongation at the translocation step, but also in reduction of the initiation rate [119]. In vitro analyses on the bacterial model have confirmed the experiments performed on the eukaryotic system, showing that ribosomes lacking adenine in the topical part of SRL are unable to stimulate GTP hydrolysis by EF-G, which is a bacterial trGTPase homologous to eukaryotic eEF2 [66]. All these experiments laid the foundation for a general notion that depurination of the SRL brings deleterious effects for the translational machinery, linking the toxic effect of ricin with blockage of protein synthesis in the cell. However, in vivo analyses have shown that there is no clear cross-correlation between the ribosome depurination, translation inhibition, and cell death [102,103], leaving the issue of ricin toxicity at the molecular level as an open question. All currently available information concerning ricin toxicity suggests that the toxic effect on cell metabolism has a multifactorial nature, involving induction of numerous pathways leading to cell death [25], but the molecular trigger is still obscure.

4. Conclusions

Ricin, in particular its catalytic subunit RTA, targets one of the most important eukaryotic ribosomal catalytic centers—the GAC, which is responsible for conferring unidirectional trajectory for the translation apparatus at the expense of GTP hydrolysis driven by trGTPases. To access the GAC, ricin hijacks the ribosomal translation factor recruitment element, i.e., the P-stalk, to reach the target adenine base in the SRL on the ribosome. The stalk, composed of P-proteins, represents a unique eukaryotic element interacting with RTA, being responsible for the specificity of the toxin toward the eukaryotic ribosome. The RTA interaction with the stalk not only anchors and directs RTA towards SRL but also stimulates depurination of invariant adenine at the tip of the SRL. The SRL represents a critical ribosomal element responsible for triggering GTP hydrolysis by trGTPases. The $G_{2659}\,A_{2660}G_{2661}A_{2662}$ (*E. coli* numbering) tetra-loop located on the tip of the SRL plays a key role here. Within this loop, two bases A_{2662} and G_{2661} are critical elements directly involved in stimulation of GTP hydrolysis [28]. On the other hand, A_{2660}, i.e., a base that is depurinated by RTA, is situated away from the trGTPase active site, being a center of cooperative interaction network, contributing to stabilization of the active state of trGTPases and promoting GTP hydrolysis. Thus, the A_{2660}, lying away from the main catalytic center, but coordinating the structural arrangement of the G domain of trGTPases, could hold the "power behind the throne" role. Therefore, depurination of solvent-exposed A_{2660} impairs the intricate interaction network and destabilizes the active state of trGTPases, which finally blocks GTP hydrolysis and, at the same time, impedes the functioning of the translational machinery. However, the majority of biochemical experiments with in vitro depurinated ribosomes have shown unequivocally that removal of A_{2660} blocks translation, especially at the elongation step of the translational cycle, linking the catalytic activity with ricin toxicity. Such a situation seems to be unusual inside the cell, where most of the toxin molecules are degraded during the ricin trafficking and, as it was estimated, only up to 5% of RTA molecules reach the endoplasmic reticulum [132–134]. Thus, ricin must have evolved to be extremely successful in winning the battle to hurt the GAC - the energetic heart of the ribosome; this is especially well demonstrated by the higher affinity towards the ribosomal P-stalk proteins [21,86,102] compared to translational factors [82]. Importantly, in vivo analyses did not find any clear link between the ribosome depurination, translation inhibition, and cell death, indicating that molecular events contributing to the so-called 'cause and effect' of the ricin *modus operandi* are still obscure.

Author Contributions: P.G. and M.T. conceived the review; P.G. and M.S. performed the literature review; P.G. and M.S. prepared the figures; P.G., M.S., P.H.-D. and M.T. wrote the manuscript.

Funding: This work was supported by the National Science Centre in Poland, grant number UMO-2014/13/B/NZ1/00953 to M.T. and UMO-2017/01/X/NZ1/01674 to P.G.

Conflicts of Interest: The authors declare no conflict of interest.

References

1.   Dang, L.; Van Damme, E.J.M. Toxic proteins in plants. *Phytochemistry* **2015**, *117*, 51–64. [CrossRef]

2.   Girbes, T.; Ferreras, J.M.; Arias, F.J.; Stirpe, F. Description, distribution, activity and phylogenetic relationship of ribosome-inactivating proteins in plants, fungi and bacteria. *Mini Rev. Med. Chem.* **2004**, *4*, 461–476. [CrossRef] [PubMed]

3. Di Maro, A.; Citores, L.; Russo, R.; Iglesias, R.; Ferreras, J.M. Sequence comparison and phylogenetic analysis by the Maximum Likelihood method of ribosome-inactivating proteins from angiosperms. *Plant. Mol. Biol.* **2014**, *85*, 575–588. [CrossRef] [PubMed]

4. Chan, T.Y. Aconitum alkaloid poisoning related to the culinary uses of aconite roots. *Toxins (Basel)* **2014**, *6*, 2605–2611. [CrossRef]

5. Barbieri, L.; Polito, L.; Bolognesi, A.; Ciani, M.; Pelosi, E.; Farini, V.; Jha, A.K.; Sharma, N.; Vivanco, J.M.; Chambery, A.; et al. Ribosome-inactivating proteins in edible plants and purification and characterization of a new ribosome-inactivating protein from Cucurbita moschata. *Biochim. Biophys. Acta.* **2006**, *1760*, 783–792. [CrossRef] [PubMed]

6. Ma, L.; Gu, R.; Tang, L.; Chen, Z.-E.; Di, R.; Long, C. Important poisonous plants in tibetan ethnomedicine. *Toxins (Basel)* **2015**, *7*, 138–155. [CrossRef]

7. Hsiao, Y.P.; Lai, W.W.; Wu, S.B.; Tsai, C.H.; Tang, S.C.; Chung, J.G.; Yang, J.H. Triggering apoptotic death of human epidermal keratinocytes by malic Acid: Involvement of endoplasmic reticulum stress- and mitochondria-dependent signaling pathways. *Toxins (Basel)* **2015**, *7*, 81–96. [CrossRef]

8. Di, R.; Tumer, N.E. Pokeweed antiviral protein: Its cytotoxicity mechanism and applications in plant disease resistance. *Toxins (Basel)* **2015**, *7*, 755–772. [CrossRef]

9. Walsh, M.J.; Dodd, J.E.; Hautbergue, G.M. Ribosome-inactivating proteins: Potent poisons and molecular tools. *Virulence* **2013**, *4*, 774–784. [CrossRef] [PubMed]

10. Christopher, G.W.; Cieslak, T.J.; Pavlin, J.A.; Eitzen, E.M., Jr. Biological warfare. A historical perspective. *JAMA* **1997**, *278*, 412–417. [CrossRef] [PubMed]

11. Audi, J.; Belson, M.; Patel, M.; Schier, J.; Osterloh, J. Ricin poisoning: A comprehensive review. *JAMA* **2005**, *294*, 2342–2351. [CrossRef]

12. Tumer, N.E.; Li, X.P. Interaction of ricin and Shiga toxins with ribosomes. *Curr. Top Microbiol. Immunol.* **2012**, *357*, 1–18. [PubMed]

13. Stirpe, F.; Barbieri, L. Ribosome-inactivating proteins up to date. *FEBS Lett.* **1986**, *195*, 1–8. [CrossRef]

14. Reinbothe, S.; Reinbothe, C.; Lehmann, J.; Becker, W.; Apel, K.; Parthier, B. JIP60, a methyl jasmonate-induced ribosome-inactivating protein involved in plant stress reactions. *Proc. Natl. Acad. Sci. USA* **1994**, *91*, 7012–7016. [CrossRef] [PubMed]

15. Hey, T.D.; Hartley, M.; Walsh, T.A. Maize ribosome-inactivating protein (b-32). Homologs in related species, effects on maize ribosomes, and modulation of activity by pro-peptide deletions. *Plant Physiol.* **1995**, *107*, 1323–1332. [CrossRef]

16. Peumans, W.J.; Van Damme, E.J.M. Evolution of Plant Ribosome-Inactivating Proteins. *Toxic Plant Proteins* **2010**, *18*, 1–26.

17. Maracci, C.; Rodnina, M.V. Review: Translational GTPases. *Biopolymers* **2016**, *105*, 463–475. [CrossRef] [PubMed]

18. Olombrada, M.; Lázaro-Gorines, R.; López-Rodríguez, J.C.; Martínez-Del-Pozo, Á.; Oñaderra, M.; Maestro-López, M.; Lacadena, J.; Gavilanes, J.G.; García-Ortega, L. Fungal Ribotoxins: A Review of Potential Biotechnological Applications. *Toxins (Basel)* **2017**, *9*, 71. [CrossRef]

19. Iglesias, R.; Citores, L.; Ragucci, S.; Russo, R.; Di Maro, A.; Ferreras, J.M. Biological and antipathogenic activities of ribosome-inactivating proteins from Phytolacca dioica L. *Biochim. Biophys. Acta.* **2016**, *1860*, 1256–1264. [CrossRef]

20. Chiou, J.C.; Li, X.P.; Remacha, M.; Ballesta, J.P.; Tumer, N.E. The ribosomal stalk is required for ribosome binding, depurination of the rRNA and cytotoxicity of ricin A chain in Saccharomyces cerevisiae. *Mol. Microbiol.* **2008**, *70*, 1441–1452. [CrossRef]

21. May, K.L.; Li, X.P.; Martínez-Azorín, F.; Ballesta, J.P.; Grela, P.; Tchórzewski, M.; Tumer, N.E. The P1/P2 proteins of the human ribosomal stalk are required for ribosome binding and depuration by ricin in human cells. *FEBS J.* **2012**, *279*, 3925–3936. [CrossRef] [PubMed]

22. Choi, A.K.; Wong, E.C.; Lee, K.M.; Wong, K.B. Structures of eukaryotic ribosomal stalk proteins and its complex with trichosanthin, and their implications in recruiting ribosome-inactivating proteins to the ribosomes. *Toxins (Basel)* **2015**, *7*, 638–647. [CrossRef] [PubMed]

23. Hudak, K.A.; Dinman, J.D.; Tumer, N.E. Pokeweed antiviral protein accesses ribosomes by binding to L3. *J. Biol. Chem.* **1999**, *274*, 3859–3864. [CrossRef] [PubMed]

24. Rajamohan, F.; Ozer, Z.; Mao, C.; Uckun, F.M. Active center cleft residues of pokeweed antiviral protein mediate its high-affinity binding to the ribosomal protein L3. *Biochemistry* **2001**, *40*, 9104–9114. [CrossRef] [PubMed]

25. Fabbrini, M.S.; Katayama, M.; Nakase, I.; Vago, R. Plant Ribosome-Inactivating Proteins: Progesses, Challenges and Biotechnological Applications (and a Few Digressions). *Toxins (Basel)* **2017**, *9*, 314. [CrossRef] [PubMed]

26. Gao, Y.G.; Selmer, M.; Dunham, C.M.; Weixlbaumer, A.; Kelley, A.C.; Ramakrishnan, V. The structure of the ribosome with elongation factor G trapped in the posttranslocational state. *Science* **2009**, *326*, 694–699. [CrossRef]

27. Clementi, N.; Polacek, N. Ribosome-associated GTPases: The role of RNA for GTPase activation. *RNA Biol.* **2010**, *7*, 521–527. [CrossRef]

28. Fischer, N.; Neumann, P.; Bock, L.V.; Maracci, C.; Wang, Z.; Paleskava, A.; Konevega, A.L.; Schröder, G.F.; Grubmüller, H.; Ficner, R.; et al. The pathway to GTPase activation of elongation factor SelB on the ribosome. *Nature* **2016**, *540*, 80–85. [CrossRef]

29. Wallin, G.; Kamerlin, S.C.; Aqvist, J. Energetics of activation of GTP hydrolysis on the ribosome. *Nat. Commun.* **2013**, *4*, 1733. [CrossRef]

30. Polikanov, Y.S.; Aleksashin, N.A.; Beckert, B.; Wilson, D.N. The Mechanisms of Action of Ribosome-Targeting Peptide Antibiotics. *Front. Mol. Biosci.* **2018**, *5*, 48. [CrossRef]

31. Wilson, D.N. Ribosome-targeting antibiotics and mechanisms of bacterial resistance. *Nat. Rev. Microbiol.* **2014**, *12*, 35–48. [CrossRef]

32. Liljas, A.; Ehrenberg, M. *Structural Aspects of Protein Synthesis*, 2nd ed.; World Scientific Publishing Company Incorporated: Hackensack, NJ, USA, 2012.

33. Nissen, P.; Hansen, J.; Ban, N.; Moore, P.B.; Steitz, T.A. The structural basis of ribosome activity in peptide bond synthesis. *Science* **2000**, *289*, 920–930. [CrossRef]

34. Rodnina, M.V.; Wintermeyer, W. The ribosome goes Nobel. *Trends Biochem. Sci.* **2010**, *35*, 1–5. [CrossRef]

35. Diaconu, M.; Kothe, U.; Schlünzen, F.; Fischer, N.; Harms, J.M.; Tonevitsky, A.G.; Stark, H.; Rodnina, M.V.; Wahl, M.C. Structural basis for the function of the ribosomal L7/12 stalk in factor binding and GTPase activation. *Cell* **2005**, *121*, 991–1004. [CrossRef]

36. Rodnina, M.V.; Stark, H.; Savelsbergh, A.; Wieden, H.J.; Mohr, D.; Matassova, N.B.; Peske, F.; Daviter, T.; Gualerzi, C.O.; Wintermeyer, W. GTPases mechanisms and functions of translation factors on the ribosome. *Biol. Chem.* **2000**, *381*, 377–387. [CrossRef] [PubMed]

37. Wahl, M.C.; Moller, W. Structure and function of the acidic ribosomal stalk proteins. *Curr. Protein Pept. Sci.* **2002**, *3*, 93–106. [CrossRef] [PubMed]

38. Voorhees, R.M.; Schmeing, T.M.; Kelley, A.C.; Ramakrishnan, V. The mechanism for activation of GTP hydrolysis on the ribosome. *Science* **2010**, *330*, 835–838. [CrossRef] [PubMed]

39. Pallesen, J.; Hashem, Y.; Korkmaz, G.; Koripella, R.K.; Huang, C.; Ehrenberg, M.; Sanyal, S.; Frank, J. Cryo-EM visualization of the ribosome in termination complex with apo-RF3 and RF1. *Elife* **2013**, *2*, e00411. [CrossRef] [PubMed]

40. Chen, Y.; Feng, S.; Kumar, V.; Ero, R.; Gao, Y.G. Structure of EF-G-ribosome complex in a pretranslocation state. *Nat. Struct. Mol. Biol.* **2013**, *20*, 1077–1084. [CrossRef]

41. Mohr, D.; Wintermeyer, W.; Rodnina, M. GTPase activation of elongation factors Tu and G on the ribosome. *Biochemistry* **2002**, *41*, 12520–12528. [CrossRef]

42. Correll, C.C.; Wool, I.G.; Munishkin, A. The two faces of the Escherichia coli 23 S rRNA sarcin/ricin domain: The structure at 1.11 A resolution. *J. Mol. Biol.* **1999**, *292*, 275–287. [CrossRef]

43. Correll, C.C.; Munishkin, A.; Chan, Y.-L.; Ren, Z.; Wool, I.G.; Steitz, T.A. Crystal structure of the ribosomal RNA domain essential for binding elongation factors. *Proc. Natl. Acad. Sci. USA* **1998**, *95*, 13436–13441. [CrossRef]

44. Ban, N.; Nissen, P.; Hansen, J.; Moore, P.B.; Steitz, T.A. The complete atomic structure of the large ribosomal subunit at 2.4 A. resolution. *Science* **2000**, *289*, 905–920. [CrossRef] [PubMed]

45. Havrila, M.; Réblová, K.; Zirbel, C.L.; Leontis, N.B.; Šponer, J. Isosteric and nonisosteric base pairs in RNA motifs: Molecular dynamics and bioinformatics study of the sarcin-ricin internal loop. *J. Phys. Chem. B* **2013**, *117*, 14302–14319. [CrossRef]

46. Yusupov, M.M.; Yusupova, G.Z.; Baucom, A.; Lieberman, K.; Earnest, T.N.; Cate, J.H.; Noller, H.F. Crystal structure of the ribosome at 5.5 A resolution. *Science* **2001**, *292*, 883–896. [CrossRef]

47. Szewczak, A.A.; Moore, P.B. The sarcin/ricin loop, a modular RNA. *J. Mol. Biol.* **1995**, *247*, 81–98. [CrossRef] [PubMed]

48. Correll, C.C.; Beneken, J.; Plantinga, M.J.; Lubbers, M.; Chan, Y.-L. The common and the distinctive features of the bulged-G motif based on a 1.04 A resolution RNA structure. *Nucleic Acids Res.* **2003**, *31*, 6806–6818. [CrossRef]

49. Heus, H.A.; Pardi, A. Structural features that give rise to the unusual stability of RNA hairpins containing GNRA loops. *Science* **1991**, *253*, 191–194. [CrossRef]

50. Moazed, D.; Robertson, J.M.; Noller, H.F. Interaction of elongation factors EF-G and EF-Tu with a conserved loop in 23S RNA. *Nature* **1988**, *334*, 362–364. [CrossRef] [PubMed]

51. Rodnina, M.V.; Savelsbergh, A.; Wintermeyer, W. Dynamics of translation on the ribosome: Molecular mechanics of translocation. *FEMS Microbiol. Rev.* **1999**, *23*, 317–333. [CrossRef]

52. Munishkin, A.; Wool, I.G. The ribosome-in-pieces: Binding of elongation factor EF-G to oligoribonucleotides that mimic the sarcin/ricin and thiostrepton domains of 23S ribosomal RNA. *Proc. Natl. Acad. Sci. USA* **1997**, *94*, 12280–12284. [CrossRef]

53. Benson, S.; Olsnes, S.; Pihl, A.; Skorve, J.; Abraham, A.K. On the mechanism of protein-synthesis inhibition by abrin and ricin. Inhibition of the GTP-hydrolysis site on the 60-S ribosomal subunit. *Eur. J. Biochem.* **1975**, *59*, 573–580. [CrossRef]

54. Sperti, S.; Montanaro, L.; Mattioli, A.; Testoni, G. Relationship between elongation factor I- and elongation factor II- dependent guanosine triphosphatase activities of ribosomes. Inhibition of both activities by ricin. *Biochem. J.* **1975**, *148*, 447–451. [CrossRef]

55. Tourigny, D.S.; Fernández, I.S.; Kelley, A.C.; Ramakrishnan, V. Elongation factor G bound to the ribosome in an intermediate state of translocation. *Science* **2013**, *340*, 1235490. [CrossRef]

56. Murray, J.; Savva, C.G.; Shin, B.S.; Dever, T.E.; Ramakrishnan, V.; Fernándezm, I.S. Structural characterization of ribosome recruitment and translocation by type IV IRES. *Elife* **2016**, *5*, e13567. [CrossRef]

57. Daviter, T.; Wieden, H.J.; Rodnina, M. Essential role of histidine 84 in elongation factor Tu for the chemical step of GTP hydrolysis on the ribosome. *J. Mol. Biol.* **2003**, *332*, 689–699. [CrossRef]

58. Maracci, C.; Peske, F.; Dannies, E.; Pohl, C.; Rodnina, M.V. Ribosome-induced tuning of GTP hydrolysis by a translational GTPase. *Proc. Natl. Acad. Sci. USA* **2014**, *111*, 14418–14423. [CrossRef]

59. Cunha, C.E.; Belardinelli, R.; Peske, F.; Holtkamp, W.; Wintermeyer, W.; Rodnina, M.V. Dual use of GTP hydrolysis by elongation factor G on the ribosome. *Translation (Austin)* **2013**, *1*, e24315. [CrossRef]

60. Adamczyk, A.J.; Warshel, A. Converting structural information into an allosteric-energy-based picture for elongation factor Tu activation by the ribosome. *Proc. Natl. Acad. Sci. USA* **2011**, *108*, 9827–9832. [CrossRef]

61. Koch, M.; Flür, S.; Kreutz, C.; Ennifar, E.; Micura, R.; Polacek, N. Role of a ribosomal RNA phosphate oxygen during the EF-G-triggered GTP hydrolysis. *Proc. Natl. Acad. Sci. USA* **2015**, *112*, E2561–E2568. [CrossRef]

62. Liljas, A.; Ehrenberg, M.; Aqvist, J. Comment on "The mechanism for activation of GTP hydrolysis on the ribosome". *Science* **2011**, *333*, 37. [CrossRef]

63. Endo, Y.; Wool, I.G. The site of action of alpha-sarcin on eukaryotic ribosomes. The sequence at the alpha-sarcin cleavage site in 28 S ribosomal ribonucleic acid. *J. Biol. Chem.* **1982**, *257*, 9054–9060. [PubMed]

64. Endo, Y.; Huber, P.W.; Wool, I.G. The ribonuclease activity of the cytotoxin alpha-sarcin. The characteristics of the enzymatic activity of alpha-sarcin with ribosomes and ribonucleic acids as substrates. *J. Biol. Chem.* **1983**, *258*, 2662–2667.

65. Grela, P.; Li, X.-P.; Horbowicz, P.; Dźwierzyńska, M.; Tchórzewski, M.; Tumer, N.E. Human ribosomal P1-P2 heterodimer represents an optimal docking site for ricin A chain with a prominent role for P1 C-terminus. *Sci. Rep.* **2017**, *7*, 5608. [CrossRef]

66. Clementi, N.; Chirkova, A.; Puffer, B.; Micura, R.; Polacek, N. Atomic mutagenesis reveals A_{2660} of 23S ribosomal RNA as key to EF-G GTPase activation. *Nat. Chem. Biol.* **2010**, *6*, 344–351. [CrossRef]

67. Avarsson, A. Structure-based sequence alignment of elongation factors Tu and G with related GTPases involved in translation. *J. Mol. Evol.* **1995**, *41*, 1096–1104. [PubMed]

68. Liljas, A.; Sanyal, S. The enigmatic ribosomal stalk. *Q Rev. Biophys.* **2018**, *51*, e12. [CrossRef] [PubMed]

69. Ballesta, J.P.; Remacha, M. The large ribosomal subunit stalk as a regulatory element of the eukaryotic translational machinery. *Prog. Nucleic Acid Res. Mol. Biol.* **1996**, *55*, 157–193. [PubMed]

70. Gonzalo, P.; Reboud, J.P. The puzzling lateral flexible stalk of the ribosome. *Biol. Cell* **2003**, *95*, 179–193. [CrossRef]

71. Davydov, I.I.; Wohlgemuth, I.; Artamonova, I.I.; Urlaub, H.; Tonevitsky, A.G.; Rodnina, M.V. Evolution of the protein stoichiometry in the L12 stalk of bacterial and organellar ribosomes. *Nat. Commun.* **2013**, *4*, 1387. [CrossRef] [PubMed]

72. Grela, P.; Gajda, M.J.; Armache, J.P.; Beckmann, R.; Krokowski, D.; Svergun, D.I.; Grankowski, N.; Tchórzewski, M. Solution structure of the natively assembled yeast ribosomal stalk determined by small-angle X-ray scattering. *Biochem. J.* **2012**, *444*, 205–209. [CrossRef] [PubMed]

73. Tchorzewski, M. The acidic ribosomal P proteins. *Int. J. Biochem. Cell Biol.* **2002**, *34*, 911–915. [CrossRef]

74. Grela, P.; Krokowski, D.; Gordiyenko, Y.; Krowarsch, D.; Robinson, C.V.; Otlewski, J.; Grankowski, N.; Tchórzewski, M. Biophysical properties of the eukaryotic ribosomal stalk. *Biochemistry* **2010**, *49*, 924–933. [CrossRef] [PubMed]

75. Krokowski, D.; Boguszewska, A.; Abramczyk, D.; Liljas, A.; Tchórzewski, M.; Grankowski, N. Yeast ribosomal P0 protein has two separate binding sites for P1/P2 proteins. *Mol. Microbiol.* **2006**, *60*, 386–400. [CrossRef]

76. Grela, P.; Sawa-Makarska, J.; Gordiyenko, Y.; Robinson, C.V.; Grankowski, N.; Tchórzewski, M. Structural properties of the human acidic ribosomal P proteins forming the P1-P2 heterocomplex. *J. Biochem.* **2008**, *143*, 169–177. [CrossRef] [PubMed]

77. Grela, P.; Bernadó, P.; Svergun, D.; Kwiatowski, J.; Abramczyk, D.; Grankowski, N.; Tchórzewski, M. Structural relationships among the ribosomal stalk proteins from the three domains of life. *J. Mol. Evol.* **2008**, *67*, 154–167. [CrossRef]

78. Lee, K.M.; Yu, C.W.; Chiu, T.Y.; Sze, K.H.; Shaw, P.C.; Wong, K.B. Solution structure of the dimerization domain of the eukaryotic stalk P1/P2 complex reveals the structural organization of eukaryotic stalk complex. *Nucleic Acids Res.* **2012**, *40*, 3172–3182. [CrossRef] [PubMed]

79. Bernado, P.; Modig, K.; Grela, P.; Svergun, D.I.; Tchorzewski, M.; Pons, M.; Akke, M. Structure and Dynamics of Ribosomal Protein L12: An Ensemble Model Based on SAXS and NMR Relaxation. *Biophys. J.* **2010**, *98*, 2374–2382. [CrossRef] [PubMed]

80. Grela, P.; Helgstrand, M.; Krokowski, D.; Boguszewska, A.; Svergun, D.; Liljas, A.; Bernadó, P.; Grankowski, N.; Akke, M.; Tchórzewski, M. Structural characterization of the ribosomal P1A-P2B protein dimer by small-angle X-ray scattering and NMR spectroscopy. *Biochemistry* **2007**, *46*, 1988–1998. [CrossRef] [PubMed]

81. Murakami, R.; Singh, C.R.; Morris, J.; Tang, L.; Harmon, I.; Takasu, A.; Miyoshi, T.; Ito, K.; Asano, K.; Uchiumi, T. The Interaction between the Ribosomal Stalk Proteins and Translation Initiation Factor 5B Promotes Translation Initiation. *Mol. Cell Biol.* **2018**, *38*. [CrossRef]

82. Tanzawa, T.; Kato, K.; Girodat, D.; Ose, T.; Kumakura, Y.; Wieden, H.J.; Uchiumi, T.; Tanaka, I.; Yao, M. The C-terminal helix of ribosomal P stalk recognizes a hydrophobic groove of elongation factor 2 in a novel fashion. *Nucleic Acids Res.* **2018**, *46*, 3232–3244. [CrossRef] [PubMed]

83. Ito, K.; Honda, T.; Suzuki, T.; Miyoshi, T.; Murakami, R.; Yao, M.; Uchiumi, T. Molecular insights into the interaction of the ribosomal stalk protein with elongation factor 1alpha. *Nucleic Acids Res.* **2014**, *42*, 14042–14052. [CrossRef] [PubMed]

84. Fan, X.; Zhu, Y.; Wang, C.; Niu, L.; Teng, M.; Li, X. Structural insights into the interaction of the ribosomal P stalk protein P2 with a type II ribosome-inactivating protein ricin. *Sci. Rep.* **2016**, *6*, 37803. [CrossRef]

85. Wawiorka, L.; Molestak, E.; Szajwaj, M.; Michalec-Wawiórka, B.; Mołoń, M.; Borkiewicz, L.; Grela, P.; Boguszewska, A.; Tchórzewski, M. Multiplication of Ribosomal P-Stalk Proteins Contributes to the Fidelity of Translation. *Mol. Cell Biol.* **2017**, *37*, e00060-17. [CrossRef] [PubMed]

86. Li, X.P.; Grela, P.; Krokowski, D.; Tchórzewski, M.; Tumer, N.E. Pentameric organization of the ribosomal stalk accelerates recruitment of ricin a chain to the ribosome for depurination. *J. Biol. Chem.* **2010**, *285*, 41463–41471. [CrossRef] [PubMed]

87. Endo, Y.; Tsurugi, K. RNA N-glycosidase activity of ricin A-chain. Mechanism of action of the toxic lectin ricin on eukaryotic ribosomes. *J. Biol. Chem.* **1987**, *262*, 8128–8130. [PubMed]

88. Agrawal, R.K.; Heagle, A.B.; Penczek, P.; Grassucci, R.A.; Frank, J. EF-G-dependent GTP hydrolysis induces translocation accompanied by large conformational changes in the 70S ribosome. *Nat. Struct. Biol.* **1999**, *6*, 643–647. [CrossRef]

89. Shi, X.; Khade, P.K.; Sanbonmatsu, K.Y.; Joseph, S. Functional role of the sarcin-ricin loop of the 23S rRNA in the elongation cycle of protein synthesis. *J. Mol. Biol.* **2012**, *419*, 125–138. [CrossRef]

90. Li, X.P.; Chiou, J.-C.; Remacha, M.; Ballesta, J.P.G.; Tumer, N.E. A two-step binding model proposed for the electrostatic interactions of ricin a chain with ribosomes. *Biochemistry* **2009**, *48*, 3853–3863. [CrossRef]

91. Endo, Y.; Tsurugi, K. The RNA N-glycosidase activity of ricin A-chain. The characteristics of the enzymatic activity of ricin A-chain with ribosomes and with rRNA. *J. Biol. Chem.* **1988**, *263*, 8735–8739.

92. Endo, Y.; Mitsui, K.; Motizuki, M.; Tsurugi, K. The mechanism of action of ricin and related toxic lectins on eukaryotic ribosomes. The site and the characteristics of the modification in 28 S ribosomal RNA caused by the toxins. *J. Biol. Chem.* **1987**, *262*, 5908–5912.

93. McCluskey, A.J.; Poon, G.M.; Bolewska-Pedyczak, E.; Srikumar, T.; Jeram, S.M.; Raught, B.; Gariépy, J. The catalytic subunit of shiga-like toxin 1 interacts with ribosomal stalk proteins and is inhibited by their conserved C-terminal domain. *J. Mol. Biol.* **2008**, *378*, 375–386. [CrossRef]

94. McCluskey, A.J.; Bolewska-Pedyczak, E.; Jarvik, N.; Chen, G.; Sidhu, S.S.; Gariépy, J. Charged and hydrophobic surfaces on the a chain of shiga-like toxin 1 recognize the C-terminal domain of ribosomal stalk proteins. *PLoS ONE* **2012**, *7*, e31191. [CrossRef]

95. Chiou, J.C.; Li, X.-P.; Remacha, M.; Ballesta, J.P.G.; Tumera, N.E. Shiga toxin 1 is more dependent on the P proteins of the ribosomal stalk for depurination activity than Shiga toxin 2. *Int. J. Biochem. Cell Biol.* **2011**, *43*, 1792–1801. [CrossRef]

96. Basu, D.; Li, X.P.; Kahn, J.N.; May, K.L.; Kahn, P.C.; Tumer, N.E. The A1 Subunit of Shiga Toxin 2 Has Higher Affinity for Ribosomes and Higher Catalytic Activity than the A1 Subunit of Shiga Toxin 1. *Infect. Immun.* **2016**, *84*, 149–161. [CrossRef]

97. Chan, S.H.; Hung, F.S.; Chan, D.S.; Shaw, P.C. Trichosanthin interacts with acidic ribosomal proteins P0 and P1 and mitotic checkpoint protein MAD2B. *Eur. J. Biochem.* **2001**, *268*, 2107–2112. [CrossRef]

98. Chan, D.S.; Chu, L.O.; Lee, K.M.; Too, P.H.; Ma, K.W.; Sze, K.H.; Zhu, G.; Shaw, P.C.; Wong, K.B. Interaction between trichosanthin, a ribosome-inactivating protein, and the ribosomal stalk protein P2 by chemical shift perturbation and mutagenesis analyses. *Nucleic Acids Res.* **2007**, *35*, 1660–1672. [CrossRef]

99. Too, P.H.; Ma, M.K.; Mak, A.N.; Wong, Y.T.; Tung, C.K.; Zhu, G.; Au, S.W.; Wong, K.B.; Shaw, P.C. The C-terminal fragment of the ribosomal P protein complexed to trichosanthin reveals the interaction between the ribosome-inactivating protein and the ribosome. *Nucleic Acids Res.* **2009**, *37*, 602–610. [CrossRef]

100. Wong, Y.T.; Ng, Y.M.; Mak, A.N.; Sze, K.H.; Wong, K.B.; Shaw, P.C. Maize ribosome-inactivating protein uses Lys158-lys161 to interact with ribosomal protein P2 and the strength of interaction is correlated to the biological activities. *PLoS ONE* **2012**, *7*, e49608. [CrossRef]

101. Li, X.P.; Kahn, P.C.; Kahn, J.N.; Grela, P.; Tumer, N.E. Arginine residues on the opposite side of the active site stimulate the catalysis of ribosome depurination by ricin A chain by interacting with the P-protein stalk. *J. Biol. Chem.* **2013**, *288*, 30270–30284. [CrossRef]

102. Grela, P.; Li, X.-P.; Tchórzewski, M.; Tumer, N.E. Functional divergence between the two P1-P2 stalk dimers on the ribosome in their interaction with ricin A chain. *Biochem. J.* **2014**, *460*, 59–67. [CrossRef] [PubMed]

103. Jetzt, A.E.; Li, X.P.; Tumer, N.E.; Cohick, W.S. Toxicity of ricin A chain is reduced in mammalian cells by inhibiting its interaction with the ribosome. *Toxicol. Appl. Pharmacol.* **2016**, *310*, 120–128. [CrossRef] [PubMed]

104. Shi, W.W.; Tang, Y.S.; Sze, S.Y.; Zhu, Z.N.; Wong, K.B.; Shaw, P.C. Crystal Structure of Ribosome-Inactivating Protein Ricin A Chain in Complex with the C-Terminal Peptide of the Ribosomal Stalk Protein P2. *Toxins (Basel)* **2016**, *8*, 296. [CrossRef]

105. Spooner, R.A.; Watson, P.D.; Marsden, C.J.; Smith, D.C.; Moore, K.A.; Cook, J.P.; Lord, J.M.; Roberts, L.M. Protein disulphide-isomerase reduces ricin to its A and B chains in the endoplasmic reticulum. *Biochem. J.* **2004**, *383 Pt 2*, 285–293. [CrossRef]

106. Endo, Y.; Gluck, A.; Wool, I.G. Ribosomal RNA identity elements for ricin A-chain recognition and catalysis. *J. Mol. Biol.* **1991**, *221*, 193–207. [CrossRef]

107. Gluck, A.; Endo, Y.; Wool, I.G. Ribosomal RNA identity elements for ricin A-chain recognition and catalysis. Analysis with tetraloop mutants. *J. Mol. Biol.* **1992**, *226*, 411–424. [CrossRef]

108. Ho, M.C.; Sturm, M.B.; Almo, S.C.; Schramm, V.L. Transition state analogues in structures of ricin and saporin ribosome-inactivating proteins. *Proc. Natl. Acad. Sci. USA* **2009**, *106*, 20276–20281. [CrossRef]

109. Jasheway, K.; Pruet, J.; Anslyn, E.V.; Robertus, J.D. Structure-based design of ricin inhibitors. *Toxins (Basel)* **2011**, *3*, 1233–1248. [CrossRef]

110. Wahome, P.G.; Robertus, J.D.; Mantis, N.J. Small-molecule inhibitors of ricin and Shiga toxins. *Curr. Top. Microbiol. Immunol.* **2012**, *357*, 179–207.

111. Kim, Y.; Robertus, J.D. Analysis of several key active site residues of ricin A chain by mutagenesis and X-ray crystallography. *Protein Eng.* **1992**, *5*, 775–779. [CrossRef]

112. Kim, Y.; Mlsna, D.; Monzingo, A.F.; Ready, M.P.; Frankel, A.; Robertus, J.D. Structure of a ricin mutant showing rescue of activity by a noncatalytic residue. *Biochemistry* **1992**, *31*, 3294–3296. [CrossRef]

113. Seggerson, K.; Moore, P.B. Structure and stability of variants of the sarcin-ricin loop of 28S rRNA: NMR studies of the prokaryotic SRL and a functional mutant. *RNA* **1998**, *4*, 1203–1215. [CrossRef]

114. Chan, Y.L.; Sitikov, A.S.; Wool, I.G. The phenotype of mutations of the base-pair C2658.G2663 that closes the tetraloop in the sarcin/ricin domain of Escherichia coli 23 S ribosomal RNA. *J. Mol. Biol.* **2000**, *298*, 795–805. [CrossRef]

115. Lin, J.Y.; Kao, W.-Y.; Tserng, K.-Y.; Chen, C.-C.; Tung, T.-C. Effect of crystalline abrin on the biosynthesis of protein, RNA, and DNA in experimental tumors. *Cancer Res.* **1970**, *30*, 2431–2433.

116. Montanaro, L.; Sperti, S.; Stirpe, F. Inhibition by ricin of protein synthesis in vitro. Ribosomes as the target of the toxin. *Biochem. J.* **1973**, *136*, 677–683. [CrossRef]

117. Olsnes, S.; Pihl, A. Ricin - a potent inhibitor of protein synthesis. *FEBS Lett.* **1972**, *20*, 327–329. [CrossRef]

118. Olsnes, S.; Fernandez-Puentes, C.; Carrasco, L.; Vazquez, D. Ribosome inactivation by the toxic lectins abrin and ricin. Kinetics of the enzymic activity of the toxin A-chains. *Eur. J. Biochem.* **1975**, *60*, 281–288. [CrossRef]

119. Osborn, R.W.; Hartley, M.R. Dual effects of the ricin A chain on protein synthesis in rabbit reticulocyte lysate. Inhibition of initiation and translocation. *Eur. J. Biochem.* **1990**, *193*, 401–407. [CrossRef]

120. Olsnes, S.; Saltvedt, E.; Pihl, A. Isolation and comparison of galactose-binding lectins from Abrus precatorius and Ricinus communis. *J. Biol. Chem.* **1974**, *249*, 803–810.

121. Endo, Y.; Tsurugi, K.; Lambert, J.M. The site of action of six different ribosome-inactivating proteins from plants on eukaryotic ribosomes: The RNA N-glycosidase activity of the proteins. *Biochem. Biophys. Res. Commun.* **1988**, *150*, 1032–1036. [CrossRef]

122. Sperti, S.; Montanaro, L.; Mattioli, A.; Stirpe, F. Inhibition by ricin of protein synthesis in vitro: 60 S ribosomal subunit as the target of the toxin. *Biochem. J.* **1973**, *136*, 813–815. [CrossRef] [PubMed]

123. Gessner, S.L.; Irvin, J.D. Inhibition of elongation factor 2-dependent translocation by the pokeweed antiviral protein and ricin. *J. Biol. Chem.* **1980**, *255*, 3251–3253.

124. Nilsson, L.; Nygard, O. The mechanism of the protein-synthesis elongation cycle in eukaryotes. Effect of ricin on the ribosomal interaction with elongation factors. *Eur. J. Biochem.* **1986**, *161*, 111–117. [CrossRef]

125. Carrasco, L.; Fernandez-Puentes, C.; Vazquez, D. Effects of ricin on the ribosomal sites involved in the interaction of the elongation factors. *Eur. J. Biochem.* **1975**, *54*, 499–503. [CrossRef]

126. Brigotti, M.; Rambelli, F.; Zamboni, M.; Montanaro, L.; Sperti, S. Effect of alpha-sarcin and ribosome-inactivating proteins on the interaction of elongation factors with ribosomes. *Biochem. J.* **1989**, *257*, 723–727. [CrossRef]

127. Nolan, R.D.; Grasmuk, H.; Drews, J. The binding of tritiated elongation-factors 1 and 2 to ribosomes from Krebs II mouse ascites-tumore cells. The influence of various antibiotics and toxins. *Eur. J. Biochem.* **1976**, *64*, 69–75. [CrossRef] [PubMed]

128. Fernandez-Puentes, C.; Benson, S.; Olsnes, S.; Pihl, A. Protective effect of elongation factor 2 on the inactivation of ribosomes by the toxic lectins abrin and ricin. *Eur. J. Biochem.* **1976**, *64*, 437–443. [CrossRef]

129. Cawley, D.B.; Hedblom, M.L.; Houston, L.L. Protection and rescue of ribosomes from the action of ricin A chain. *Biochemistry* **1979**, *18*, 2648–2654. [CrossRef]

130. Sperti, S.; Montanaro, L. Ricin and modeccin do not inhibit the elongation factor 1-dependent binding of aminoacyl-tRNA to ribosomes. *Biochem. J.* **1979**, *178*, 233–236. [CrossRef] [PubMed]

131. Montanaro, L.; Sperti, S.; Mattioli, A.; Testoni, G.; Stirpe, F. Inhibition by ricin of protein synthesis in vitro. Inhibition of the binding of elongation factor 2 and of adenosine diphosphate-ribosylated elongation factor 2 to ribosomes. *Biochem. J.* **1975**, *146*, 127–131. [CrossRef] [PubMed]

132. Spooner, R.A.; Lord, J.M. How ricin and Shiga toxin reach the cytosol of target cells: Retrotranslocation from the endoplasmic reticulum. *Curr. Top Microbiol. Immunol.* **2012**, *357*, 19–40. [PubMed]

133. van Deurs, B.; Sandvig, K.; Petersen, O.W.; Olsnes, S.; Simons, K.; Griffiths, G. Estimation of the amount of internalized ricin that reaches the trans-Golgi network. *J. Cell. Biol.* **1988**, *106*, 253–267. [CrossRef] [PubMed]
134. van Deurs, B.; Petersen, O.W.; Olsnes, S.; Sandvig, K. Delivery of internalized ricin from endosomes to cisternal Golgi elements is a discontinuous, temperature-sensitive process. *Exp. Cell. Res.* **1987**, *171*, 137–152. [CrossRef]

Article

Peptide Mimics of the Ribosomal P Stalk Inhibit the Activity of Ricin A Chain by Preventing Ribosome Binding

Xiao-Ping Li, Jennifer N. Kahn and Nilgun E. Tumer *

Department of Plant Biology, School of Environmental and Biological Sciences, Rutgers, The State University of New Jersey, New Brunswick, NJ 08901-8520, USA; xpli@sebs.rutgers.edu (X.-P.L.); jennifer.nielsen.kahn@gmail.com (J.N.K.)
* Correspondence: tumer@sebs.rutgers.edu; Tel.: +1-848-932-6359

Received: 10 August 2018; Accepted: 10 September 2018; Published: 13 September 2018

Abstract: Ricin A chain (RTA) depurinates the sarcin/ricin loop (SRL) by interacting with the C-termini of the ribosomal P stalk. The ribosome interaction site and the active site are located on opposite faces of RTA. The interaction with P proteins allows RTA to depurinate the SRL on the ribosome at physiological pH with an extremely high activity by orienting the active site towards the SRL. Therefore, if an inhibitor disrupts RTA–ribosome interaction by binding to the ribosome binding site of RTA, it should inhibit the depurination activity. To test this model, we synthesized peptides mimicking the last 3 to 11 amino acids of P proteins and examined their interaction with wild-type RTA and ribosome binding mutants by Biacore. We measured the inhibitory activity of these peptides on RTA-mediated depurination of yeast and rat liver ribosomes. We found that the peptides interacted with the ribosome binding site of RTA and inhibited depurination activity by disrupting RTA–ribosome interactions. The shortest peptide that could interact with RTA and inhibit its activity was four amino acids in length. RTA activity was inhibited by disrupting its interaction with the P stalk without targeting the active site, establishing the ribosome binding site as a new target for inhibitor discovery.

Keywords: ricin A chain; ribosomal P stalk; P protein interaction; SRL depurination; peptide inhibition

Key Contribution: Peptides that bind to the ribosome binding site of RTA can inhibit the depurination activity by disrupting RTA–ribosome interactions, demonstrating that the ribosome binding site is a potential new target for inhibitor development.

---

1. Introduction

Ricin (E.C. 3.2.2.22), produced by the castor bean (*Ricinus communis*), belongs to a group of toxic proteins called ribosome-inactivating proteins (RIPs) that include major human pathogens, such as *Escherichia coli* and *Shigella* producing Shiga toxins (Stxs). They are classified as category B agents of national security and public health risk with potential for significant morbidity and mortality. Currently, no U.S. Food and Drug Administration-approved vaccines or therapeutics exist to protect against ricin, Stxs, or any other RIP. The RIPs cleave a universally conserved adenine from the sarcin/ricin loop (SRL) on the large rRNA, inhibiting protein synthesis and inducing cell death [1–4]. Ricin A chain (RTA) interacts with the P proteins of the ribosomal stalk to depurinate the SRL [5]. Several other RIPs, including Shiga toxins [6–8], also interact with the conserved C-terminal domain (CTD) of P proteins in order to access the SRL [9–11].

In eukaryotes, the P stalk is a pentameric protein complex composed of two P1/P2 dimers that bind to the C-terminus of the uL10 (previously P0) protein, while the N-terminal domain of uL10 anchors the stalk on the large subunit of the ribosome [12–14]. The P stalk and the SRL are part of the GTPase-associated center (GAC) in the large subunit [13,15]. The unique feature of all P proteins is the 11 C-terminal amino acids, which are identical in all eukaryotes and have a disordered structure [16]. The CTD of P proteins selectively recognizes translational GTPases, such as the initiation factor 5B (eIF5B) and the elongation factors eEF-2/EFG and eEF1α/EFTu, and recruits them to the SRL [17–20].

The interaction of RTA with P proteins is critical for ribosome binding, depurination of the SRL, and toxicity of RTA in the yeast *Saccharomyces cerevisiae* and in human cells [5,21]. We showed that the ribosome binding site and the active site are located on opposite faces of RTA and based on these results we proposed a molecular model for depurination of the SRL by RTA [22]. According to this model, RTA is concentrated around the ribosome by electrostatic interactions [23]. The P stalk interacts with RTA and stalk binding stimulates the catalysis of depurination by orienting the active site of RTA towards the SRL [22]. The interaction with the P proteins allows RTA to depurinate the SRL on the ribosome at physiological pH with an extremely high catalytic activity, while at this pH RTA is not active on the naked RNA [22,24]. Based on this model, we predict that if an inhibitor disrupts the interaction of RTA with the ribosome by binding to the ribosome binding site of RTA, it should be able to inhibit the depurination activity of RTA.

Until now, no inhibitor targeting the ribosome binding site of RTA has been reported. A 17-mer peptide mimicking the CTD of the human ribosomal stalk P2 protein was shown to inhibit the activity of the A1 subunit of Shiga toxin 1 (Stx1A1) in an in vitro translation assay [6]. Calmodulin-tagged peptides corresponding to the last 11 and 17 residues of the human P2 protein could pull down Stx1A1 and RTA, but not a peptide corresponding to the last 7 residues of human P2 [6]. The 11-mer peptide (P11), $S_{105}DDDMGFGLFD_{115}$, contains a negatively charged acidic motif "$D_{106}D_{107}D_{108}$" at its N-terminus and a hydrophobic "$F_{111}GLFD_{115}$" motif at its C-terminus. The pull down and binding studies showed that both motifs are important for the interaction of Stx1A1 with P11 [7].

In *S. cerevisiae*, P1/P2 proteins have diverged into P1α/P2β and P1β/P2α [25]. The P1α/P2β dimer could inhibit the depurination activity of Stx1A1 and Stx2A1 on ribosomes isolated from a yeast mutant in which the binding site of the P protein dimers on the P0 protein had been deleted, suggesting that P1α/P2β can bind to the toxins and prevent them from depurinating ribosomes [8]. Although the amino acid sequences of the A1 subunit of Stx1 and Stx2 are only 21% and 20% identical to RTA, respectively, they are structurally and functionally very similar to RTA [26]. Recently, the structure of RTA together with the last six amino acids of P proteins was resolved by two different groups using different strategies (PDB IDs: 5GU4 and 5DDZ) [27,28]. Although 9-mer, 11-mer [27], or 10-mer [28] P stalk peptides were used in the studies, the structures resolved by both groups showed that only the last six residues interact with RTA. The last six residues ($G_{110}FGLFD_{115}$) bind to a hydrophobic pocket on RTA in a unique conformation. Neither group could visualize the N-terminal end of the peptide that contains the negatively charged ($D_{106}D_{107}D_{108}$) motif. However, the two groups drew different conclusions. One group postulated that this motif contributes to the interaction because GST-tagged peptides, which contained the $D_{106}D_{107}D_{108}$ motif, showed higher affinities for RTA than those without this motif [27]. In contrast, the other group concluded that the $D_{106}D_{107}D_{108}$ motif does not contribute to the RTA interaction because they did not observe any difference in the pull-down experiments with His-tagged RTA when this motif was mutated or deleted [28]. We previously showed that arginines at the RTA/RTB interface contribute to fast electrostatic interactions with the CTD of the P proteins, indicating that the negatively charged motif plays an important role in the interaction of RTA with the ribosome [29].

Although the active site of RTA has been explored extensively as a target for antidotes, the interaction of RTA with ribosomes has not been previously examined as a potential drug target. To better understand the recognition mechanism of the P protein CTD by RTA and to define the minimal length of a peptide that can bind RTA and inhibit its activity, we measured the interaction of peptides corresponding to the last 3 to 11 amino acids of human P proteins with RTA and examined their ability to inhibit the depurination activity of RTA. We discuss the relationship between the affinity of the peptides and their inhibitory activity. Our results establish the ribosome binding site of RTA as a new target for inhibitor discovery. Since Stxs also bind to the P protein CTD to depurinate the SRL, a similar approach could be explored for Stxs.

2. Results

2.1. The Longer Peptides Have Higher Affinity for RTA than the Shorter Peptides

The sequences of peptides mimicking the last 11 amino acids of the P proteins are shown in Figure 1.

P11	S_{105}	D_{106}	D_{107}	D_{108}	M_{109}	G_{110}	F_{111}	G_{112}	L_{113}	F_{114} D_{115}
P10		D_{106}	D_{107}	D_{108}	M_{109}	G_{110}	F_{111}	G_{112}	L_{113}	F_{114} D_{115}
P9			D_{107}	D_{108}	M_{109}	G_{110}	F_{111}	G_{112}	L_{113}	F_{114} D_{115}
P8				D_{108}	M_{109}	G_{110}	F_{111}	G_{112}	L_{113}	F_{114} D_{115}
P7					M_{109}	G_{110}	F_{111}	G_{112}	L_{113}	F_{114} D_{115}
P6						G_{110}	F_{111}	G_{112}	L_{113}	F_{114} D_{115}
P5							F_{111}	G_{112}	L_{113}	F_{114} D_{115}
P4								G_{112}	L_{113}	F_{114} D_{115}
P3									L_{113}	F_{114} D_{115}

Figure 1. The sequence of the peptides corresponding to the C-terminal end of P proteins.

We used surface plasmon resonance (SPR) with Biacore T200 (GE Healthcare, Marlborough, MA, USA) to measure the affinity of these peptides for RTA. Because of solubility limitations, the highest concentration of peptide measured was 0.5 mM. For direct comparison of affinity with depurination activity, depurination buffer was used as the running buffer for the Biacore analysis. The dose-dependent interaction sensorgrams are shown in Figure 2a and the fitting is shown in Figure 2b. The interaction sensorgrams of peptides with RTA showed fast on and fast off characteristics, suggesting fast association and fast dissociation, and thus relatively low affinity (Figure 2b). As the peptide concentration increased, the equilibrium binding levels increased. Due to the fast association and dissociation, equilibrium data were used to calculate the dissociation constants (K_D). The K_Ds of the peptides are shown in Table 1. Overall, the affinity of the peptides for RTA was in the high micromolar range. The affinity of RTA for the peptides was 10^5 times lower than its affinity for the ribosome, which is in the low nanomolar range [23]. The K_D increased from 196 μM to 451 μM as the number of amino acids decreased from 11 (P11) to 4 (P4). Deletion of the three aspartic acids ($D_{106}D_{107}D_{108}$) did not decrease the affinity dramatically, as the K_D was 272 μM for P10 and around 300 μM for P9, P8, and P7. The decreased affinity was mainly due to the deletion of Asp106. The K_D did not change much when Asp107 and Asp108 were deleted. The K_D increased from 294 μM to 399 μM with the deletion of Met109 and to 497 μM when Gly110 was deleted, indicating the importance of Met109 and Gly110 in the interaction. Deleting Phe111 did not change the affinity appreciably. However, upon deleting Gly112, the affinity decreased dramatically, indicating that the peptide with only the last three amino acids of P proteins almost lost the ability to bind RTA.

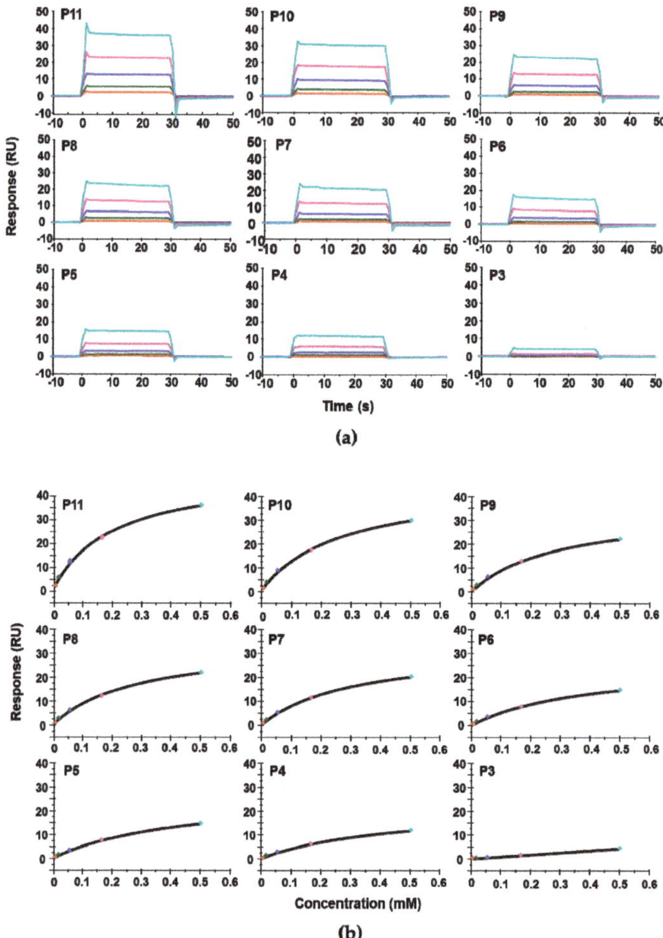

Figure 2. The interaction curves (**a**) and the steady state affinity fitting (**b**) of the peptide—Ricin A chain (RTA) interaction. The K_D was determined using Biacore T200. The untagged recombinant RTA was immobilized on a CM5 chip by amine coupling at about 2422 RU. The reference surface was activated and blocked. The peptides were passed over the surface at 6.2 μM (red), 18.5 μM (green), 55.6 μM (dark blue), 166.7 μM (magenta), and 500 μM (light blue).

Table 1. The affinity of peptides for RTA.

Peptides	Sequence	MW	K_D (μM) *
P11	S_{105}DDDMGFGLFD$_{115}$	1218.25	196 ± 17
P10	D_{106}DDMGFGLFD$_{115}$	1131.18	272 ± 6
P9	D_{107}DMGFGLFD$_{115}$	1016.09	309 ± 7
P8	D_{108}MGFGLFD$_{115}$	901.00	299 ± 5
P7	M_{109}GFGLFD$_{115}$	785.91	294 ± 47
P6	G_{110}FGLFD$_{115}$	654.71	399 ± 20
P5	F_{111}GLFD$_{115}$	597.66	497 ± 30
P4	G_{112}LFD$_{115}$	450.49	451 ± 17
P3	L_{113}FD$_{115}$	393.44	>10 mM

* Data were obtained from the fitting shown in Figure 1 and shown as average ± SD of three to four replicates. MW: molecular weight. K_D: the equilibrium dissociation constant.

2.2. Peptides Bind to the Ribosome Binding Site of RTA

We showed that the R189A, R193A, R234A, and R235A mutations affect the ribosome binding of RTA [29]. The order of decreased binding strength was R235A > R234A > R193A $\geq$ R189A [29], where the R235A mutant completely lost ribosome binding [29]. To verify the binding site of these peptides on RTA, the His-tagged RTA mutants R189A, R193A, R234A, and R235A were captured on an NTA chip and P11 and P4 were passed over the surface. As shown in Figure 3, the binding levels of P11 and P4 decreased for the Arg mutants. The relative decrease in binding levels of the Arg mutants to P11 and P4 was in the same order as the relative decrease in ribosome binding. The lowest binding was observed with R235A, followed by R234A, R193A, and R189A, indicating that the peptides bind at the ribosome binding site of RTA.

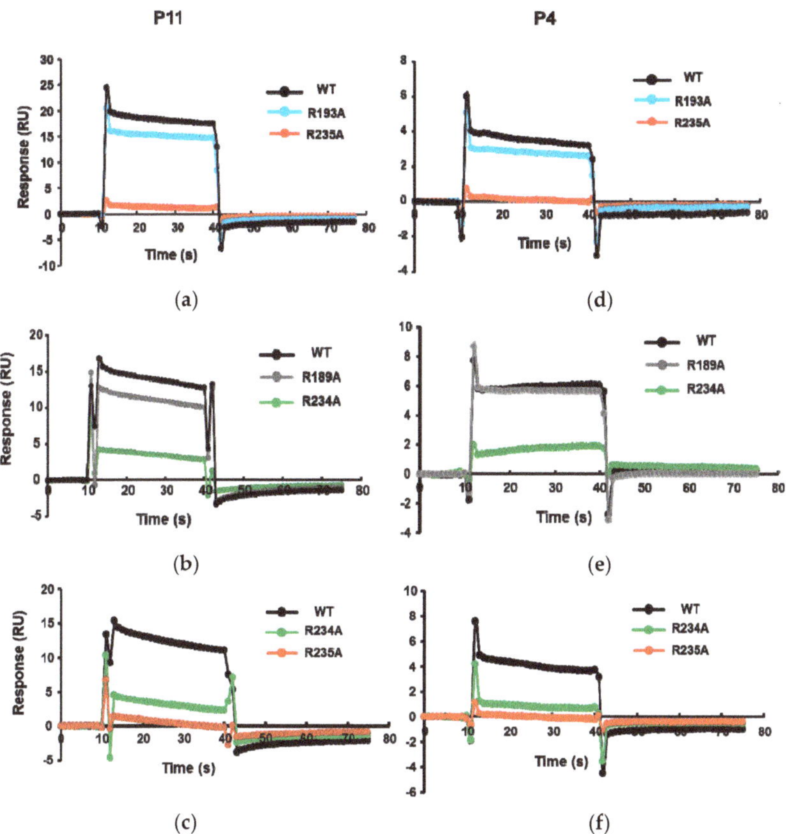

Figure 3. The interaction curves of P11 (**a–c**) and P4 (**d–f**) with wild-type (WT) RTA and the R193A, R235A, R189A, and R234A point mutants. The interactions were analyzed by Biacore T200 using the same conditions as in Figure 2, except N-terminally His-tagged wild-type RTA (10×His-RTA) or 10×His-tagged RTA mutants were captured on an NTA chip at around 2100 RU. Flow cell 1 (Fc1) was used as control, R235A or R234A was captured on Fc2, R193A or R189A was captured on Fc3, and wild type (WT)-RTA was captured on Fc4. The P11 or P4 were passed over the surface at a concentration of 166.7 µM. The signals were normalized to the chip density of WT-RTA.

The crystal structures of P6 with RTA have shown that Arg235 and the last aspartic acid of P proteins, Asp115, form a critical hydrogen bond. To determine if Asp115 is critical for binding RTA, a 5-mer peptide P5b ($G_{110}FGLF_{114}$) in which Asp115 was deleted was synthesized and its interaction with RTA was compared to P5 ($F_{111}GLFD_{115}$). The calculated K_D of P5b was over 20 times higher than that of P5, indicating that the last aspartic acid (D_{115}) is critical for the interaction with RTA (Figure 4). These results provide further evidence that the peptides interact with the ribosome binding site of RTA.

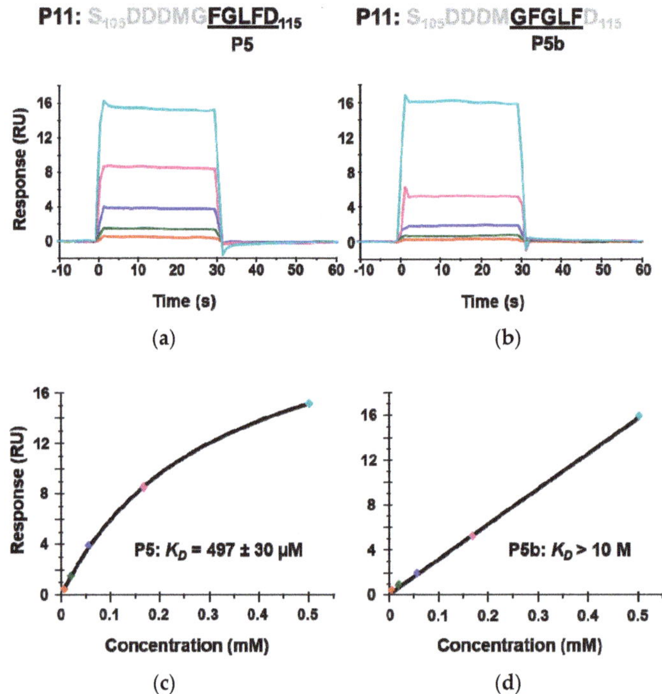

Figure 4. The interaction of peptides mimicking the last five amino acids (P5) of the P proteins (**a,c**) and the penultimate five amino acids (P5b) with RTA (**b,d**). The K_D was determined by Biacore T200 using the same conditions as in Figure 2. The binding sensorgrams are shown in (**a,b**) and the fitting is shown in (**c,d**).

2.3. Peptides Compete with Ribosomes for Binding to RTA

To determine if peptide binding at the ribosome binding site of RTA can affect binding of RTA to ribosomes, a peptide-ribosome competition assay was conducted using the A-B-A injection capability of Biacore 8K. The molecular weight of the yeast ribosome is about 3000 times larger than that of the peptides. The affinity of RTA for the ribosome is in the nanomolar range [23]. The affinity of peptides for RTA is in the high micromolar range (Table 1) and about 10^5-fold lower than the affinity for the ribosome. Since the structure of the last six amino acids of P proteins with RTA has been resolved, P6 was chosen for the competition assay. RTA was immobilized on a CM5 chip by amine coupling. The surface was first blocked by P6 at three different concentrations (125, 250, and 500 µM) for one minute. Ribosomes (20 nM) were mixed with the same three concentrations of P6 and passed over the surface for two minutes. The ribosome binding levels at the end of each injection were plotted against the concentration of P6. As shown in Figure 5, ribosome binding decreased as the concentration of P6 increased, indicating that P6 competed with ribosomes for binding to RTA in a dose-dependent manner. However, 10^4-fold higher concentration of P6 was needed compared to the ribosome and the inhibition did not reach 100%.

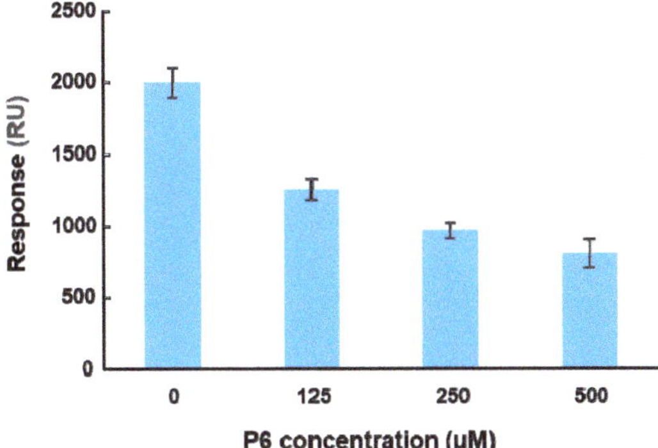

Figure 5. P6 competes with the ribosome for binding to RTA. The A-B-A capability of Biacore 8K was used for the competition analysis. RTA was immobilized on a CM5 chip at 4000 RU by amine coupling. The P6 was passed over the RTA at indicated concentrations for 1 min and then yeast ribosomes (20 nM) were injected over the surface together with the same concentrations of P6 for another 2 min at a flow rate of 30 µL per min. The ribosome binding levels were determined at 5 s before the end of the injection. The surface was regenerated by three one-minute injections of 2 M NaCl and one injection of running buffer with 2% DMSO. The data are expressed as average ± SD from four replicates.

2.4. Peptides Inhibit the Depurination Activity of RTA

We showed that the peptides bind to the ribosome binding site of RTA and affect the interaction of RTA with the ribosome. Based on our ribosome depurination model, if ribosome binding of RTA is reduced then depurination of the SRL on the ribosome should be affected. To test this model, we examined the ability of the peptides to inhibit the depurination activity of RTA on yeast and rat liver ribosomes. Both ribosomes were used since RTA depurinates rat liver ribosomes at a much higher rate than yeast ribosomes [30]. Different concentrations of peptides were incubated with RTA first, and then ribosomes were added to start the depurination reaction. After 5 min at room temperature, the reaction was stopped and the level of ribosome depurination was measured by qRT-PCR [31]. The percent depurination was plotted against the concentration of the peptide (Figure 6). The IC_{50} values were obtained by fitting the inhibition curves with the Michaelis–Menten equation using the Origin Pro 9.1 software (OriginLab, Northampton, MA, USA). The inhibition of depurination of yeast (Figure 6a) and rat liver (Figure 6b) ribosomes by P11 are shown. As peptide concentrations increased, the percentage of inhibition increased, and percent inhibition reached 90% for both ribosomes. The IC_{50} of P11 was 4.7 µM for yeast ribosomes and 31 µM for rat liver ribosomes (Table 2). The IC_{50} values of P10 to P3 were measured using the same method for yeast (Figure S1) and rat liver (Figure S2) ribosomes and data are shown in Table 2. The IC_{50} values for yeast ribosomes increased from 7.9 µM to 15 µM when D_{106} was deleted and to 23 µM and 34 µM when D_{107} and D_{108} were deleted, respectively. Similarly, the IC_{50} values for rat ribosomes increased from 83 µM to 142 µM when D_{106} was deleted and to 267 µM when D_{107} was deleted. These results demonstrated that the $D_{106}D_{107}D_{108}$ motif is important for inhibition of the depurination activity of RTA. The IC_{50} values were about 6 to 10 times higher for rat ribosomes than yeast ribosomes, possibly because RTA depurinates rat ribosomes at a higher rate than yeast ribosomes [30]. We could not measure IC_{50} for P7 to P4 for rat ribosomes because the peptide concentrations needed were prohibitive and limited by solubility. We could not detect any inhibition activity for P3 and P5b at the highest concentration measured (500 µM). The IC_{50} values correlated with the RTA interaction results (Table 1) and indicated that inhibition of ribosome binding by peptides with even a low affinity could lead to inhibition of the depurination activity of RTA.

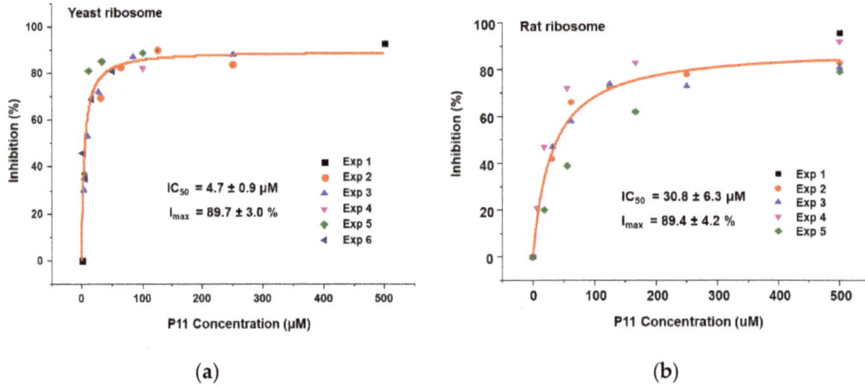

Figure 6. Inhibition of depuration activity of RTA by P11. The depurination levels were determined by qRT-PCR. (**a**) Yeast ribosomes were used at 60 nM and RTA was used at 1.0 nM. (**b**) Rat liver ribosomes were used at 60 nM and RTA was used at 0.2 nM. Different concentrations of P11 and RTA were mixed first and the reaction was started by adding ribosomes. The reaction was incubated at the room temperature for 5 min and was stopped by adding 2×RNA extraction buffer. The RNA was purified and the depurination levels were determined by qRT-PCR. The depurination level of the reaction without toxin was set as 100%. The depurination levels were calculated and plotted as percent of no toxin control. Experiments were conducted four to six times and the data were fit with the Michaelis–Menten equation using Origin Pro 9.1. IC_{50}: the half maximal inhibitory concentration. I_{max}: maximal inhibition.

Table 2. Inhibition of depurination activity of RTA by peptides.

Peptide	Sequence	Yeast Ribosome IC_{50} (μM) *	Rat Ribosome IC_{50} (μM) *
P11	S_{105}DDDMGFGLFD$_{115}$	4.7 ± 0.9	31 ± 6.3
P10	D_{106}DDMGFGLFD$_{115}$	7.9 ± 1.7	83 ± 18
P9	D_{107}DMGFGLFD$_{115}$	15 ± 1.5	142 ± 61
P8	D_{108}MGFGLFD$_{115}$	23 ± 4.4	267 ± 80
P7	M_{109}GFGLFD$_{115}$	34 ± 9.5	NA
P6	G_{110}FGLFD$_{115}$	63 ± 13	NA
P5	F_{111}GLFD$_{115}$	121 ± 44	NA
P4	G_{112}LFD$_{115}$	102 ± 45	NA

* The IC_{50} values were determined using the method shown in Figure 6 and Figures S1 and S2. Data are from the fitting results. NA: not analyzed.

3. Discussion

The active site of RTA has been explored extensively as a potential target for antidotes against depurination [32]. However, since SRL is the substrate of RTA, the active site is large and mostly polar and therefore small molecule inhibitor screens have yielded few potential inhibitors with low affinity [32]. Although inhibitors showed activity in enzymatic tests, they failed to protect cells or animals against ricin challenge. Only one small molecule has been shown to have activity in protecting mice against ricin challenge by blocking the retrograde trafficking of ricin [33]. To address this barrier and to establish a starting point for inhibitor discovery, we established the ribosome binding site of RTA as a new target and identified the shortest length of a peptide that can bind to the ribosome binding site of RTA and inhibit its activity.

The C-terminal ends of the ribosomal stalk P proteins interact with a small well-defined hydrophobic pocket on the face of RTA opposite to the active site [27,28]. We show here that peptides derived from the conserved CTD of P proteins can disrupt RTA–ribosome interactions. The longest peptide tested was P11 because this is the smallest peptide reported to inhibit the activity of Stx1 [6].

Since the C-terminal end of this peptide is critical for RTA interaction [27,28], we deleted one amino acid at a time starting from the N-terminal end. We found that the longer peptides had higher affinity and inhibitory activity compared to the shorter peptides. The shortest peptide that could interact with RTA and inhibit its activity corresponded to the last four amino acids of P proteins.

A conserved $D_{106}D_{107}D_{108}$ motif at the N-terminal end of P11 has been shown to be critical for the interaction with trichosanthin (TCS) [34]. In the structure of the TCS–P11 complex, Asp108 of P11 interacted with Lys173 of TCS via salt bridges, while Asp106 of P11 formed hydrogen bonds with Gln169 of TCS [35]. However, in the two structures of this peptide with RTA, the $D_{106}D_{107}D_{108}$ residues were not observed [27,28]. The substitution of Asp108 and Asp106 residues in P2 with alanine abolished the interaction between P2 and TCS [35]. However, these mutations did not affect the interaction of P2 with RTA [28]. Based on these results, Fan et al. concluded that the conserved $D_{106}D_{107}D_{108}M_{109}$ motif of P2 is not involved in the interaction with RTA and only hydrophobic interactions and hydrogen bonds contribute to the interaction [28]. In contrast, Shi et al. showed that although the $D_{106}D_{107}D_{108}$ motif was not observed in the structure, the binding affinity of RTA measured by isothermal calorimetry (ITC) was lower when this motif was not present on the peptide than when this motif was present [27].

To address the role of the negatively charged motif at the P protein CTD, we deleted these residues one at a time. Our results indicate that individual deletion of Asp_{107} and Asp_{108} in the $D_{106}D_{107}D_{108}$ motif had negligible effect on affinity of the peptide for RTA, while deletion of Asp_{106} had a small effect. This may explain why the GST-tagged P2 variants containing Asp to Ala mutations in this motif interacted with RTA in the pull-down experiments [28]. When affinity was measured directly using Biacore T200, we observed a small decrease in K_D from 272 µM to about 300 µM when Asp_{106} was deleted. However, the IC_{50} increased 2-fold upon deletion of Asp_{106}. The IC_{50} increased further 2-fold when Asp_{107} was deleted (Table 2). This data is consistent with our previous study [29] where mutation of positively charged arginines on RTA led to a significant increase in K_m toward ribosomes without affecting the K_m or k_{cat} towards an RNA mimic of the SRL, indicating that electrostatic contacts contribute to the interaction of RTA with the ribosome [29]. We showed that arginines are critical for maintaining the fast association and dissociation rates of the interaction with the CTD of P proteins [29]. We proposed that these arginines form a positively charged patch on the surface of RTA and interact with the negatively charged $D_{106}D_{107}D_{108}$ motif at the CTD of the P proteins to facilitate the interaction of RTA with the P stalk to allow depurination of the SRL [29]. The results presented here provide direct evidence that the $D_{106}D_{107}D_{108}$ motif is important for ribosome anchoring of RTA for depurination of the SRL.

The qRT-PCR based depurination assay, which directly measures the catalytic activity of RTA on ribosomes showed demonstrable impact on RTAs ability to depurinate ribosomes in a manner which correlated with ribosome binding [36]. Comparison of the binding affinity and the IC_{50} of RTA for the peptides indicated that the IC_{50} values were lower than the K_D values. The P11 with 200 µM K_D was able to achieve 50% inhibition of RTA activity at 5 µM, which is about 40 times lower than the K_D. For the shorter peptides, such as P5 and P4, the IC_{50} (~100 µM) values were about 5 times lower than the K_D (~500 µM) values. Although peptides do not bind RTA tightly, as indicated by the high K_D values, they inhibit the activity of RTA, as indicated by the relatively low IC_{50} values. This difference may be because of allosteric binding sites, where peptides are binding in a location separate from the active site. Thus, peptides do not compete with binding of the active site of RTA to the SRL, but compete with binding of RTA to the P stalk. Our results indicate that inhibition of depurination activity involves both electrostatic and hydrophobic surfaces on the P protein peptide. Electrostatic interactions are critical to maintain the high association and dissociation rates of RTA with the P proteins on the ribosome [28]. Even low affinity binding to the P protein peptide led to a high level of inhibition of depurination by RTA, suggesting that the ribosome binding site is a potentially valuable target distinct from the active site. However, due to the low affinity of the peptides for RTA in combination with the high catalytic efficiency of RTA the therapeutic potential of the peptides used in this study is limited and is not a claim of this paper. They need to be optimized into higher affinity ligands. We establish the ribosome binding site as a potential new target for inhibitor discovery as a proof of concept. Since P protein CTD is the binding site of several RIPs, including the Stxs, inhibitors targeting the ribosome interaction site of RTA could be effective against the Stxs.

The structural analysis showed that the binding between P10 and RTA is mediated by hydrophobic interactions. The Phe111, Leu 113, and Phe114 residues are inserted into a hydrophobic pocket and the Phe114 and Asp115 residues form hydrogen bonds with Arg235 of RTA [27,28]. Consistent with the structural analysis, our results indicated that P5 and P4 showed similar affinity and inhibition activity for RTA (Tables 1 and 2). The binding and depurination inhibition results of P5 over P5b (Figure 4) confirmed that Asp115 plays an important role in the interaction. However, P3 containing both Phe114 and Asp115 only bound very weakly to RTA and could not inhibit RTA activity. Although the conformation of TCS bound to P11 differed from the structure of RTA, both RIPs recognized the Leu113 and Phe114 motif [28]. This LF motif is conserved in both eukaryotic and archaeal ribosomal stalk proteins and has been shown to be necessary for binding of translational GTPases to the stalk proteins [37,38]. Our results suggest that RTA binding to P3 is substantially reduced when Gly112 is deleted possibly because Gly112 accommodates the required backbone to facilitate the insertion of Phe111, Leu113, and Phe114 into the hydrophobic pocket on RTA [28]. The structural analysis showed that the C-terminal sequences of P proteins do not form a stable structure in solution in a ligand-free state [16], but appear as an α-helix upon binding to the hydrophobic pocket of RTA [27,28]. Since 3.6 amino acids are the minimum length needed to form an α-helix, the last three amino acids may not interact well with RTA because they cannot form an α-helix even though they contain all the critical side chains for the interaction. These results indicate that the minimal length of P protein CTD required for binding to RTA and inhibiting its activity is four amino acids.

The C-terminal sequences of the stalk proteins can adapt diverse conformations in order to bind distinct ligands specifically [16]. The P11 appeared as a type II β-turn upon binding to TCS [35]. However, the last six amino acids of P11 formed an α-helix when bound to RTA [27,28]. Similarly, the CTD of archaeal P1 (aP1) formed a β-turn and a 3_{10}-helix when bound to eIF5B [20]. In contrast, the CTD of aP1 bound to eIF1A formed a long extended α-helix [37]. The aP1 bound to a hydrophobic pocket on the surface of eIF5B, which is present on the opposite side of the GTP/GDP binding site [20]. It was suggested that the stalk/eIF5B interaction contributes to the recruitment of the GTP binding site of eIF5B to the SRL [20]. Our results indicate that RTA interacts with rapid on and off rates and with low affinity with peptides mimicking the C-terminal sequence of the P proteins. The A1 subunit of Stx1 was also shown to interact with low affinity with P11 [7]. RIPs and translational GTPases may interact with low affinity with the stalk proteins to properly orient their active site and the GTP binding site, respectively, towards the SRL. Thus, our RTA–ribosome interaction model, which proposes that the interaction with the stalk stimulates the catalysis of depurination by orienting the active site of RTA towards the SRL [22], may be applicable to other RIPs and the translational GTPases that interact with the stalk.

4. Materials and Methods

4.1. Peptide Synthesis

Peptides were synthesized by GenScript (Piscataway, NJ, USA) at purity higher than 98%. The purity and the sequence accuracy were confirmed by HPLC and mass spectrometry.

4.2. RTA Purification

Wild-type untagged RTA was purified using a previously published method [30]. His-tagged wild-type RTA was purified as previously described [22]. His-tagged RTA mutants were purified by the Northeast Biodefense Center Protein Expression Core. Yeast and rat liver ribosomes were purified using a previously published method [5].

4.3. Peptide–RTA Interaction

The peptide–RTA interaction was measured by surface plasmon resonance (SPR) using Biacore T200. The untagged RTA was immobilized on a CM5 chip at 2000 to 3000 RU by amine coupling. The reference channel was blocked as the active channel. The peptides were passed over the surface at 6.2, 18.5, 55.6, 166.7, and 500 µM at a flow rate of 30 µL/min. The running buffer was 10 mM Tris-HCl pH7.5, 60 mM KCl, 10 mM MgOAc, and 0.5% DMSO. The affinity was obtained by fitting the binding data using Biacore T200 Evaluation Software version 3.0.

The peptide–RTA mutant interactions were measured using Biacore T200. The 10×His-tagged wild-type RTA or the 10×His-tagged RTA mutants were captured on an NTA chip to around 2000 RU. The blank surface without Ni^+ was used as the reference. The peptides were passed over the surface at 30 µL/min. The running buffer was the same as the untagged RTA. The binding signals were normalized for the differences in surface density.

4.4. Peptide Competition Assays

The peptide competition assays were conducted using Biacore 8K (GE Healthcare, Marlborough, MA, USA). The untagged RTA was immobilized on a CM5 chip to around 4000 RU in flow cell 2 for all eight channels by amine coupling. The flow cell 1 was activated and blocked. The A-B-A injection was used. The surface was first injected for 1 minute with peptide P6 at 0, 125, 250, and 500 µM. Then, yeast ribosomes (20 nM) mixed with the same concentrations of P6 were injected on the surface for 2 min. Ribosome disassociation was allowed for another 2 min. The flow rate was 30 µL/min. The surface was regenerated by three one-minute injections of 2 M NaCl and one 1 min injection of running buffer containing 2% DMSO. The running buffer was the same as peptide–RTA interaction buffer with 0.005% of surfactant P20. The binding levels of ribosome at "A-B-A binding later" were compared.

4.5. Inhibition of RTA Depurination

The inhibition of depurination activity was measured by qRT-PCR using the untagged RTA [31]. The final RTA concentration was 1.0 nM for yeast ribosomes and 0.2 nM for rat liver ribosomes with both ribosome concentrations at 60 nM. The same buffer used to examine the interaction of RTA with peptides was used. The final peptide concentrations varied dependent on the level of inhibition by the peptides. RTA was mixed with the peptides at room temperature for several minutes. The ribosomes were added to start the reaction. The depurination reaction was incubated at 25 °C for 5 min and was stopped by adding an equal volume of RNA extraction buffer. RNA was purified from the depurinated ribosomes using a previously published method [22]. The percentage of depurination was determined by qRT-PCR [31]. In each set of measurements, the ribosome and peptide mixture without toxin was used as 100% and the mixture with ribosome and toxin but without the peptide was used as 0%. The percentage of inhibition was plotted against the peptide concentration and the IC_{50} was calculated by fitting the data with the Michaelis–Menten function using Origin Pro 9.1.

Supplementary Materials: The following are available online at http://www.mdpi.com/2072-6651/10/9/371/s1, Figure S1: Inhibition of depurination activity of RTA on yeast ribosomes by peptides P10–P3. The IC_{50} values were determined as in Figure 6 and are shown in Table 2. Figure S2: Inhibition of depurination activity of RTA on rat liver ribosomes by peptides P10, P9, and P8. The IC_{50} values were determined as in Figure 6 and are shown in Table 2.

Author Contributions: X.-P.L., J.N.K. and N.E.T. planned experiments, X.-P.L. and J.N.K. performed experiments, X.-P.L., J.N.K. and N.E.T. interpreted data, X.-P.L. and N.E.T. wrote the paper. All authors reviewed the paper.

Funding: This work was supported by National Institutes of Health grants AI072425 and AI127980 to Nilgun E. Tumer.

Acknowledgments: We thank Yijun Zhou for critical reading of the paper. We thank Karen Chave (Northeast Biodefense Center Protein Synthesis Core, U54-AI057158-Lipkin) for purification of RTA variants from *E. coli*.

Conflicts of Interest: The authors declare no conflict of interest.

References

1. Bolognesi, A.; Bortolotti, M.; Maiello, S.; Battelli, M.G.; Polito, L. Ribosome-inactivating proteins from plants: A historical overview. *Molecules* **2016**, *21*, 1627. [CrossRef] [PubMed]
2. Olsnes, S. The history of ricin, abrin and related toxins. *Toxicon* **2004**, *44*, 361–370. [CrossRef] [PubMed]
3. Stirpe, F. Ribosome-inactivating proteins. *Toxicon* **2004**, *44*, 371–383. [CrossRef] [PubMed]
4. Tesh, V.L.; O'Brien, A.D. The pathogenic mechanisms of Shiga toxin and the Shiga-like toxins. *Mol. Microbiol.* **1991**, *5*, 1817–1822. [CrossRef] [PubMed]
5. Chiou, J.C.; Li, X.P.; Remacha, M.; Ballesta, J.P.; Tumer, N.E. The ribosomal stalk is required for ribosome binding, depurination of the rrna and cytotoxicity of ricin A chain in saccharomyces cerevisiae. *Mol. Microbiol.* **2008**, *70*, 1441–1452. [CrossRef] [PubMed]
6. McCluskey, A.J.; Poon, G.M.; Bolewska-Pedyczak, E.; Srikumar, T.; Jeram, S.M.; Raught, B.; Gariepy, J. The catalytic subunit of Shiga-like toxin 1 interacts with ribosomal stalk proteins and is inhibited by their conserved C-terminal domain. *J. Mol. Biol.* **2008**, *378*, 375–386. [CrossRef] [PubMed]
7. McCluskey, A.J.; Bolewska-Pedyczak, E.; Jarvik, N.; Chen, G.; Sidhu, S.S.; Gariepy, J. Charged and hydrophobic surfaces on the a chain of Shiga-like toxin 1 recognize the C-terminal domain of ribosomal stalk proteins. *PLoS ONE* **2012**, *7*, e31191. [CrossRef] [PubMed]
8. Chiou, J.C.; Li, X.P.; Remacha, M.; Ballesta, J.P.; Tumer, N.E. Shiga toxin 1 is more dependent on the P proteins of the ribosomal stalk for depurination activity than Shiga toxin 2. *Int. J. Biochem. Cell Biol.* **2011**, *43*, 1792–1801. [CrossRef] [PubMed]
9. Shi, W.W.; Mak, A.N.; Wong, K.B.; Shaw, P.C. Structures and ribosomal interaction of ribosome-inactivating proteins. *Molecules* **2016**, *21*, 1588. [CrossRef] [PubMed]
10. Choi, A.K.; Wong, E.C.; Lee, K.M.; Wong, K.B. Structures of eukaryotic ribosomal stalk proteins and its complex with trichosanthin, and their implications in recruiting ribosome-inactivating proteins to the ribosomes. *Toxins* **2015**, *7*, 638–647. [CrossRef] [PubMed]
11. Tumer, N.E.; Li, X.P. Interaction of ricin and Shiga toxins with ribosomes. *Curr. Top. Microbiol. Immunol.* **2012**, *357*, 1–18. [PubMed]
12. Jose, M.P.; Santana-Roman, H.; Remacha, M.; Ballesta, J.P.; Zinker, S. Eukaryotic acidic phosphoproteins interact with the ribosome through their amino-terminal domain. *Biochemistry* **1995**, *34*, 7941–7948. [CrossRef] [PubMed]
13. Tchorzewski, M. The acidic ribosomal P proteins. *Int. J. Biochem. Cell Biol.* **2002**, *34*, 911–915. [CrossRef]
14. Krokowski, D.; Boguszewska, A.; Abramczyk, D.; Liljas, A.; Tchorzewski, M.; Grankowski, N. Yeast ribosomal P0 protein has two separate binding sites for P1/P2 proteins. *Mol. Microbiol.* **2006**, *60*, 386–400. [CrossRef] [PubMed]
15. Wahl, M.C.; Moller, W. Structure and function of the acidic ribosomal stalk proteins. *Curr. Protein Pept. Sci.* **2002**, *3*, 93–106. [CrossRef] [PubMed]
16. Lee, K.M.; Yusa, K.; Chu, L.O.; Yu, C.W.; Oono, M.; Miyoshi, T.; Ito, K.; Shaw, P.C.; Wong, K.B.; Uchiumi, T. Solution structure of human P1*P2 heterodimer provides insights into the role of eukaryotic stalk in recruiting the ribosome-inactivating protein trichosanthin to the ribosome. *Nucleic Acids Res.* **2013**, *41*, 8776–8787. [CrossRef] [PubMed]
17. Villa, E.; Sengupta, J.; Trabuco, L.G.; Lebarron, J.; Baxter, W.T.; Shaikh, T.R.; Grassucci, R.A.; Nissen, P.; Ehrenberg, M.; Schulten, K.; et al. Ribosome-induced changes in elongation factor Tu conformation control GTP hydrolysis. *Proc. Natl. Acad. Sci. USA* **2009**, *106*, 1063–1068. [CrossRef] [PubMed]
18. Schmeing, T.M.; Voorhees, R.M.; Kelley, A.C.; Gao, Y.G.; Murphy, F.V.T.; Weir, J.R.; Ramakrishnan, V. The crystal structure of the ribosome bound to EF-Tu and aminoacyl-tRNA. *Science* **2009**, *326*, 688–694. [CrossRef] [PubMed]
19. Tanzawa, T.; Kato, K.; Girodat, D.; Ose, T.; Kumakura, Y.; Wieden, H.J.; Uchiumi, T.; Tanaka, I.; Yao, M. The C-terminal helix of ribosomal P stalk recognizes a hydrophobic groove of elongation factor 2 in a novel fashion. *Nucleic Acids Res.* **2018**, *46*, 3232–3244. [CrossRef] [PubMed]
20. Murakami, R.; Singh, C.R.; Morris, J.; Tang, L.; Harmon, I.; Azuma, T.; Miyoshi, T.; Ito, K.; Asano, K.; Uchiumi, T. The interaction between the ribosomal stalk proteins and translation initiation factor 5B promotes translation initiation. *Mol. Cell. Biol.* **2018**, *38*, e00067-18. [CrossRef] [PubMed]

21. May, K.L.; Li, X.P.; Martinez-Azorin, F.; Ballesta, J.P.; Grela, P.; Tchorzewski, M.; Tumer, N.E. The P1/P2 proteins of the human ribosomal stalk are required for ribosome binding and depurination by ricin in human cells. *FEBS J.* **2012**, *279*, 3925–3936. [CrossRef] [PubMed]

22. Li, X.P.; Kahn, P.C.; Kahn, J.N.; Grela, P.; Tumer, N.E. Arginine residues on the opposite side of the active site stimulate the catalysis of ribosome depurination by ricin A chain by interacting with the P-protein stalk. *J. Biol. Chem.* **2013**, *288*, 30270–30284. [CrossRef] [PubMed]

23. Li, X.P.; Chiou, J.C.; Remacha, M.; Ballesta, J.P.; Tumer, N.E. A two-step binding model proposed for the electrostatic interactions of ricin A chain with ribosomes. *Biochemistry* **2009**, *48*, 3853–3863. [CrossRef] [PubMed]

24. Chen, X.Y.; Link, T.M.; Schramm, V.L. Ricin A-chain: Kinetics, mechanism, and RNA stem-loop inhibitors. *Biochemistry* **1998**, *37*, 11605–11613. [CrossRef] [PubMed]

25. Krokowski, D.; Tchorzewski, M.; Boguszewska, A.; Grankowski, N. Acquisition of a stable structure by yeast ribosomal P0 protein requires binding of P1a-P2b complex: In vitro formation of the stalk structure. *Biochim. Biophys. Acta* **2005**, *1724*, 59–70. [CrossRef] [PubMed]

26. Li, X.P.; Tumer, N.E. Differences in ribosome binding and sarcin/ricin loop depurination by Shiga and ricin holotoxins. *Toxins* **2017**, *9*, 133. [CrossRef] [PubMed]

27. Shi, W.W.; Tang, Y.S.; Sze, S.Y.; Zhu, Z.N.; Wong, K.B.; Shaw, P.C. Crystal structure of ribosome-inactivating protein ricin A chain in complex with the C-terminal peptide of the ribosomal stalk protein P2. *Toxins* **2016**, *8*, 296. [CrossRef] [PubMed]

28. Fan, X.; Zhu, Y.; Wang, C.; Niu, L.; Teng, M.; Li, X. Structural insights into the interaction of the ribosomal P stalk protein P2 with a type II ribosome-inactivating protein ricin. *Sci. Rep.* **2016**, *6*, 37803. [CrossRef] [PubMed]

29. Zhou, Y.; Li, X.P.; Chen, B.Y.; Tumer, N.E. Ricin uses arginine 235 as an anchor residue to bind to P-proteins of the ribosomal stalk. *Sci. Rep.* **2017**, *7*, 42912. [CrossRef] [PubMed]

30. Sturm, M.B.; Schramm, V.L. Detecting ricin: Sensitive luminescent assay for ricin A-chain ribosome depurination kinetics. *Anal. Chem.* **2009**, *81*, 2847–2853. [CrossRef] [PubMed]

31. Pierce, M.; Kahn, J.N.; Chiou, J.; Tumer, N.E. Development of a quantitative RT-PCR assay to examine the kinetics of ribosome depurination by ribosome inactivating proteins using *Saccharomyces cerevisiae* as a model. *RNA* **2011**, *17*, 201–210. [CrossRef] [PubMed]

32. Wahome, P.G.; Robertus, J.D.; Mantis, N.J. Small-molecule inhibitors of ricin and Shiga toxins. *Curr. Top. Microbiol.* **2012**, *357*, 179–207.

33. Stechmann, B.; Bai, S.K.; Gobbo, E.; Lopez, R.; Merer, G.; Pinchard, S.; Panigai, L.; Tenza, D.; Raposo, G.; Beaumelle, B.; et al. Inhibition of retrograde transport protects mice from lethal ricin challenge. *Cell* **2010**, *141*, 231–242. [CrossRef] [PubMed]

34. Chan, D.S.; Chu, L.O.; Lee, K.M.; Too, P.H.; Ma, K.W.; Sze, K.H.; Zhu, G.; Shaw, P.C.; Wong, K.B. Interaction between trichosanthin, a ribosome-inactivating protein, and the ribosomal stalk protein P2 by chemical shift perturbation and mutagenesis analyses. *Nucleic Acids Res.* **2007**, *35*, 1660–1672. [CrossRef] [PubMed]

35. Too, P.H.M.; Ma, M.K.W.; Mak, A.N.S.; Wong, Y.T.; Tung, C.K.C.; Zhu, G.; Au, S.W.N.; Wong, K.B.; Shaw, P.C. The C-terminal fragment of the ribosomal P protein complexed to trichosanthin reveals the interaction between the ribosome-inactivating protein and the ribosome. *Nucleic Acids Res.* **2009**, *37*, 602–610. [CrossRef] [PubMed]

36. Zhou, Y.; Li, X.-P.; Kahn, J.N.; Tumer, N.E. Functional assays for measuring the catalytic activity of ribosome inactivating proteins. *Toxins* **2018**, *10*, 240. [CrossRef] [PubMed]

37. Ito, K.; Honda, T.; Suzuki, T.; Miyoshi, T.; Murakami, R.; Yao, M.; Uchiumi, T. Molecular insights into the interaction of the ribosomal stalk protein with elongation factor 1α. *Nucleic Acids Res.* **2014**, *42*, 14042–14052. [CrossRef] [PubMed]

38. Nomura, N.; Honda, T.; Baba, K.; Naganuma, T.; Tanzawa, T.; Arisaka, F.; Noda, M.; Uchiyama, S.; Tanaka, I.; Yao, M.; et al. Archaeal ribosomal stalk protein interacts with translation factors in a nucleotide-independent manner via its conserved C terminus. *Proc. Natl. Acad. Sci. USA* **2012**, *109*, 3748–3753. [CrossRef] [PubMed]

Article

TNF Family Cytokines Induce Distinct Cell Death Modalities in the A549 Human Lung Epithelial Cell Line when Administered in Combination with Ricin Toxin

Alexa L. Hodges [1,†], Cody G. Kempen [1,†], William D. McCaig [1], Cory A. Parker [1], Nicholas J. Mantis [2,*] and Timothy J. LaRocca [1,*]

[1] Department of Basic and Clinical Sciences, Albany College of Pharmacy and Health Sciences, Albany, NY 12208, USA
[2] Division of Infectious Disease, Wadsworth Center, New York State Department of Health, Albany, NY 12208, USA
* Correspondence: Timothy.LaRocca@acphs.edu (T.J.L.); nicholas.mantis@health.ny.gov (N.J.M.)
† These authors contributed equally.

Received: 19 June 2019; Accepted: 28 July 2019; Published: 1 August 2019

Abstract: Ricin is a member of the ribosome-inactivating protein (RIP) family of toxins and is classified as a biothreat agent by the Centers for Disease Control and Prevention (CDC). Inhalation, the most potent route of toxicity, triggers an acute respiratory distress-like syndrome that coincides with near complete destruction of the lung epithelium. We previously demonstrated that the TNF-related apoptosis-inducing ligand (TRAIL; CD253) sensitizes human lung epithelial cells to ricin-induced death. Here, we report that ricin/TRAIL-mediated cell death occurs via apoptosis and involves caspases -3, -7, -8, and -9, but not caspase-6. In addition, we show that two other TNF family members, TNF-α and Fas ligand (FasL), also sensitize human lung epithelial cells to ricin-induced death. While ricin/TNF-α- and ricin/FasL-mediated killing of A549 cells was inhibited by the pan-caspase inhibitor, zVAD-fmk, evidence suggests that these pathways were not caspase-dependent apoptosis. We also ruled out necroptosis and pyroptosis. Rather, the combination of ricin plus TNF-α or FasL induced cathepsin-dependent cell death, as evidenced by the use of several pharmacologic inhibitors. We postulate that the effects of zVAD-fmk were due to the molecule's known off-target effects on cathepsin activity. This work demonstrates that ricin-induced lung epithelial cell killing occurs by distinct cell death pathways dependent on the presence of different sensitizing cytokines, TRAIL, TNF-α, or FasL.

Keywords: ricin; toxins; cytokines; toxin-mediated diseases; apoptosis; cathepsin; tumor necrosis factor; fas; caspases

Key Contribution: In this study, we have defined the cell death pathways induced by ricin/TRAIL, ricin/TNF-α, and ricin/FasL in human lung epithelial cells. As our analysis probed each pathway at several steps, this work may lead to novel therapeutic approaches to ricin toxicity that target multiple cell death pathways at different steps in addition to direct targeting of ricin.

---

1. Introduction

Ricin is a 60–65 kDa glycoprotein toxin derived from the castor bean plant, *Ricinus communis* [1–3]. The toxin, which presumably functions in plant defense, comprises 1–5% of the total dry weight of the bean. The cytotoxicity of ricin is based on its ability to inhibit protein synthesis in all mammalian cell types, including macrophages and epithelial cells [1–3]. Due to its potential to be aerosolized and deployed as a biological weapon, ricin is classified by the Centers for Disease Control and Prevention (CDC) as a select agent [1–5].

Technically, ricin is a member of the type II family of ribosome-inactivating proteins (RIPs), consisting of a catalytic A subunit (RTA) attached via disulfide bond to a cell-binding B subunit (RTB) [6,7]. RTB is a lectin that binds β-1,4 galactose (Gal) and N-acetylgalactosamine (GalNAc) moieties on glycolipids and glycoproteins on the surface of target cells [8]. Following binding, ricin is internalized via clathrin-dependent endocytosis and then undergoes retrograde transport to the trans Golgi network (TGN) and endoplasmic reticulum (ER) [9]. Recently, it has been shown that fucosylation and the absence of sialylation are vital for the trafficking of ricin to these compartments [9]. Once the toxin reaches the ER, the disulfide link between the A and B subunits is reduced and RTA alone is translocated into the cell cytoplasm [9,10]. RTA inhibits protein synthesis by virtue of its ability to cleave a specific glycosidic bond in the so-called sarcin-ricin loop (SRL) of rRNA in the 60s ribosomal subunit [10–12]. The SRL is critical for the binding of elongation factor 2 to the ribosome, which is necessary for polypeptide synthesis [13,14]. Therefore, depurination of the SRL leads to the cessation of protein synthesis [11,12]. This activity is so potent that it has been noted that a single RTA can inhibit the function of 1500 ribosomes per minute [1]. Ricin is extremely toxic following inhalation [15,16]. Wide-scale damage caused by inhaled ricin leads to acute respiratory distress syndrome (ARDS) which is characterized by a potent proinflammatory response [16–19].

Previously, we reported that the cytokine TNF-α related apoptosis-inducing ligand (TRAIL) modulates the toxicity of ricin as well as the host inflammatory response to this toxin [20]. In particular, we demonstrated that addition of TRAIL enhanced the death of Calu-3 human lung epithelial cells in a caspase-dependent manner and evoked an inflammatory response dominated by IL-6 [20]. Considering that TRAIL is one of a number of potent cell death ligands that accumulate during proinflammatory responses [21–23], we wanted to evaluate the cell death modulatory activities of other cytokines in the context of ricin toxicity. These cytokines include TNF-α and Fas ligand (FasL), both of which, along with TRAIL, are capable of inducing several different programmed cell death pathways [21–23]. In addition, proinflammatory and death-inducing cytokines such as these are abundant components in the bronchoalveolar lavage fluid of animals following ricin inhalation [24–29]. We hypothesize that lung epithelial cells compromised by ricin will be primed to undergo high levels of cell death following contact with death-inducing cytokines. We believe that this heightened cell death response to ricin will be controlled by known programmed cell death pathways. In the current study, we use biochemical approaches to provide a detailed characterization of A549 human lung epithelial cell death responses to ricin administered in combination with TRAIL, TNF-α, or FasL. Defining these cell death responses and identifying multiple steps at which they can be inhibited may lead to new therapeutic approaches to ricin toxicity targeted against specific programmed cell death pathways.

2. Results

2.1. Ricin-Induced Death Is Primed by TRAIL, TNF-α, and FasL

To test the hypothesis that extrinsic cytokines cause an increase in ricin-induced cell death, A549 cells were treated with increasing doses of ricin in the absence or presence of 100 ng/mL TRAIL, TNF-α, or FasL for 24 h at 37 °C. Over a range of toxin concentrations, addition of each of the cytokines resulted in a significant increase in ricin-induced cell death (Figure 1A). Our previous work indicated that ricin/TRAIL induces apoptosis of Calu-3 human lung epithelial cells [20]. Indeed, when we used the pan-caspase inhibitor zVAD-fmk, A549 cell death induced by ricin combined with TRAIL, TNF-α, or FasL was prevented (Figure 1B–D). Collectively, the results of Figure 1 indicate that TRAIL, TNF-α, and FasL enhance ricin-induced cell death in a manner that is likely caspase-dependent apoptosis.

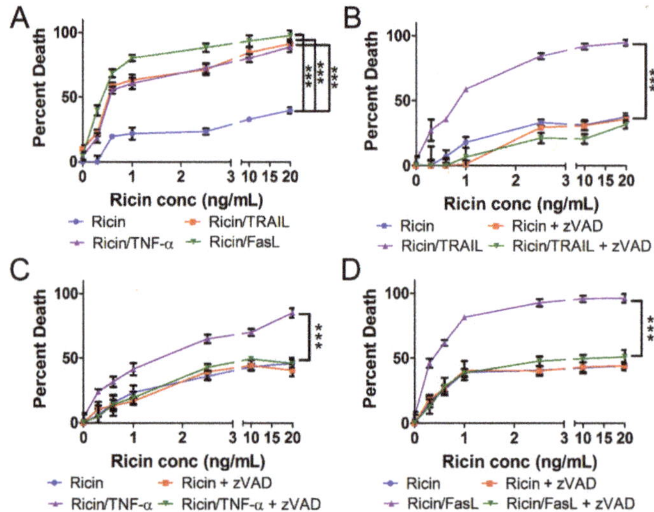

Figure 1. Extrinsic cytokines enhance ricin-induced death of A549 cells in a zVAD inhibitable manner. A549 lung epithelial cells were treated with ricin alone or in combination with 100 ng/mL TRAIL, TNF-α, or FasL for 24h at 37 °C followed by measurement of cell death via WST-1 assay. (**A**) All 3 cytokines/ligands resulted in a significant increase in cell death relative to treatment with ricin alone. A549 cell death induced by ricin combined with (**B**) TRAIL, (**C**) TNF-α, or (**D**) FasL is prevented by the pan-caspase inhibitor, zVAD-fmk (50 µM). Results are the average of 3 independent experiments. Error bars = standard deviation. Two-way analysis of variance (ANOVA), *** $p < 0.001$.

2.2. Cell Death Induced by Ricin/TRAIL Is Associated with Caspase Activation while Death by Ricin/TNF-α and Ricin/FasL Is Not

To get a clear view of the involvement of caspases, the effectors of apoptosis [21], in cell death by ricin combined with TRAIL, TNF-α, or FasL, A549 cells were treated with 1 ng/mL ricin combined with 100 ng/mL TRAIL, TNF-α, or FasL for 4 h at 37 °C followed by cell lysis and western blot. Treatment with ricin or any of the cytokines alone did not result in caspase cleavage/activation (Figure 2A–C). When combined with TRAIL, ricin induced cleavage/activation of caspases-3, -7, -8, and -9 but not caspase-6 (Figure 2A and Figure S1). However, the combination of ricin and TNF-α or ricin and FasL did not cause cleavage/activation of any caspase tested (Figure 2B,C). To determine if caspases were cleaved with slower kinetics, we measured caspase cleavage in response to ricin/TNF-α or ricin/FasL after 8 h of treatment. However, caspases were not cleaved/activated in A549 cells at this time point (Figure S2). As a positive control for TRAIL-, TNF-, and FasL-induced apoptosis, A549 cells were treated with 250 ng/mL cycloheximide (CHX) combined with 100 ng/mL TRAIL, TNF-α, or FasL [30–37]. Apoptosis induced by CHX combined with any of the cytokines resulted in cleavage/activation of caspases-3, -6, -7, -8, and -9 (Figure 2D and Figure S3). These results clearly demonstrate that caspases are activated following attack by ricin/TRAIL but are not affected by ricin/TNF-α and ricin/FasL.

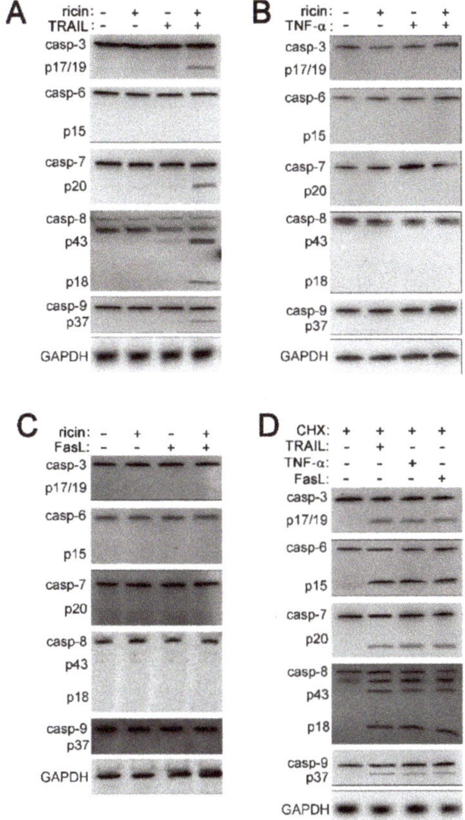

Figure 2. The combination of ricin and TRAIL induces caspase activation, while addition of TNF-α or FasL with ricin does not. A549 lung epithelial cells were treated with 1 ng/mL ricin alone or in combination with 100 ng/mL TRAIL, TNF-α, or FasL for 4 h at 37 °C followed by cell lysis and western blot. (**A**) The combination of ricin and TRAIL results in cleavage/activation of caspases-3, -7, -8, and -9. Caspase-6 is not cleaved/activated in response to ricin/TRAIL (**B,C**) The combination of ricin/TNF-α or ricin/FasL does not result in the cleavage/activation of caspases. (**D**) A549 cells were treated with 250 ng/mL cycloheximide (CHX) combined with TNF-α, FasL, or TRAIL as a positive control for TRAIL-, TNF-, and FasL-induced apoptosis. As expected, when cycloheximide is combined with TNF-α, FasL, or TRAIL it results in the cleavage/activation of caspases-3, -6, -7, -8, and -9. Shown are representative blots from 3 independent experiments.

2.3. The Combination of Ricin and TRAIL Induces Caspase-Dependent Apoptosis

The results of Figure 2 suggest that ricin/TRAIL causes activation of caspases, and thus apoptosis, while ricin/TNF-α and ricin/FasL do not. This finding was unexpected for ricin/TNF-α and ricin/FasL. Therefore, we decided to perform a detailed characterization of A549 cell death caused by ricin combined with TRAIL, TNF-α, or FasL. A549 cells were treated with increasing doses of ricin and 100 ng/mL TRAIL for 24 h at 37 °C in the absence or presence of specific caspase inhibitors. We determined that cell death induced by ricin combined with TRAIL depends on executioner caspases-3 and -7 (Figure 3A and Figure S4) as well as initiator caspases-8 and -9 (Figure 3C,D and Figure S4) but not caspase-6 (Figure 3B). These results are in agreement with our caspase activation/cleavage results of Figure 2. As a positive control for TRAIL-induced apoptosis, A549 cells were treated with

250 ng/mL CHX combined with increasing concentrations of TRAIL [30] (Figure 3E). Cell death induced by CHX/TRAIL was prevented by inhibition of caspases-3, -7, -8, and -9 as the combination of CHX and TRAIL induces apoptosis [30,38,39]. These results indicate that ricin induces caspase-dependent apoptosis of human lung epithelial cells when combined with TRAIL similar to the combination of CHX and TRAIL.

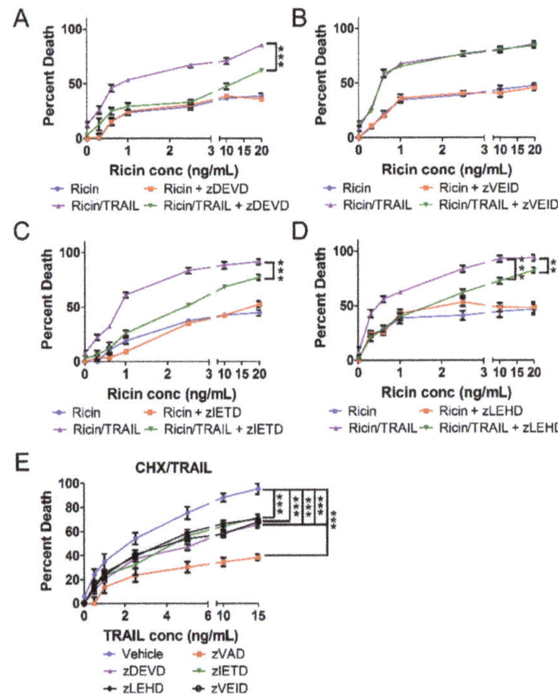

Figure 3. A549 cell death induced by ricin/TRAIL depends on caspases-3, -7, -8, and -9. A549 lung epithelial cells were treated with ricin alone or in combination with 100 ng/mL TRAIL in the presence or absence of caspase inhibitors for 24 h at 37 °C followed by measurement of cell death via WST-1 assay. Cell death induced by the combination of ricin and TRAIL was prevented by inhibition of (**A**) caspases-3 and -7 with zDEVD-fmk (30 μM) but not (**B**) caspase-6 with zVEID-fmk (30 μM). Moreover, cell death by ricin/TRAIL was prevented by inhibitions of (**C**) caspase-8 with zIETD-fmk (30 μM) and (**D**) caspase-9 with zLEHD-fmk (10 μM). (**E**) A549 cells were treated with the combination of 250 ng/mL cycloheximide (CHX) and TRAIL as a positive control for TRAIL-induced apoptosis. As expected, cycloheximide/TRAIL-induced apoptosis is prevented by all caspase inhibitors tested. Results are the average of 3 independent experiments. Error bars = standard deviation. Two-way ANOVA, *** $p < 0.001$, ** $p < 0.01$.

2.4. Ricin Combined with TNF-α or FasL Induces Caspase-Independent Cell Death

We next characterized A549 cell death by ricin/TNF-α or ricin/FasL with respect to the findings of Figure 2. A549 cells were treated with increasing doses of ricin and 100 ng/mL TNF-α or FasL for 24 h at 37 °C. Cell death induced by ricin/TNF-α or ricin/FasL was not prevented by specific inhibition of caspases-3, -6, -7, -8, or -9 (Figure 4A–F and Figure S5). In contrast to the results with zVAD-fmk (Figure 1C,D), these results are in agreement with those of Figure 2. As positive controls for TNF- and FasL-induced apoptosis, A549 cells were treated with 250 ng/mL CHX combined with increasing concentrations of TNF-α [30–37] (Figure 4G) or FasL [30,34,36] (Figure 4H). TNF- and FasL-induced apoptosis were prevented by inhibition of caspases-3, -7, -8, and -9 (Figure 4G,H). These results indicate that ricin/TNF-α and ricin/FasL induce caspase-independent cell death which is distinct from the caspase-dependent apoptosis induced by CHX/TNF-α and CHX/FasL.

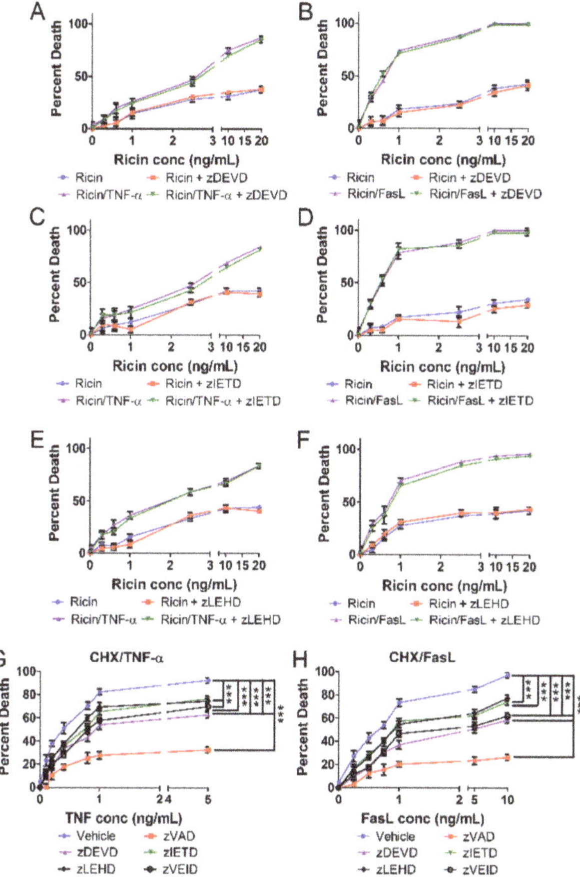

Figure 4. Ricin/TNF-α and ricin/FasL both induce caspase-independent death of A549 cells. A549 lung epithelial cells were treated with ricin alone or in combination with 100 ng/mL TNF-α or FasL in the presence or absence of caspase inhibitors for 24 h at 37 °C followed by measurement of cell death via WST-1 assay. Cell death induced by ricin/TNF-α and ricin/FasL was not prevented by inhibition of (**A,B**) caspases-3 and -7 with zDEVD-fmk (30 μM), (**C,D**) caspase-8 with zIETD-fmk (30 μM), or (**E,F**) caspase-9 with zLEHD-fmk (10 μM) (**G,H**) A549 cells were treated with the combination of 250 ng/mL cycloheximide (CHX) and TNF-α or cycloheximide and FasL as positive controls for TNF- and FasL-induced apoptosis. As expected, cycloheximide/TNF- and cycloheximide/FasL-induced apoptosis is prevented by all caspase inhibitors tested. Results are the average of 3 independent experiments. Error bars = standard deviation. Two-way ANOVA, *** $p < 0.001$.

Since cell death induced by ricin/TNF-α and ricin/FasL was prevented by the pan-caspase inhibitor, zVAD-fmk but not inhibition of specific apoptotic caspases (Figures 1C,D and 4 and Figure S5), we investigated the involvement of alternative caspases and apoptosis effectors in both instances of cell death. Inhibition of caspase-1, the central effector caspase of pyroptosis [40,41], did not prevent cell death by ricin/TNF-α or ricin/FasL (Figure 5A,B). Caspase-2 is thought to have initiator and effector roles in some forms of apoptosis [42] yet its inhibition did not prevent cell death by ricin combined with either cytokine (Figure 5C,D). During intrinsic apoptosis, Bax is critical for mitochondrial outer membrane pore formation and cytochrome *c* release [21]. However, when Bax was inhibited there was no effect on the cell death induced by ricin/TNF-α or ricin/FasL (Figure 5E,F) in contrast to its effect on apoptosis induced by CHX/TNF-α or CHX/FasL (Figure S6). Furthermore, cytochrome *c* was not observed in the cytoplasm of cells treated with ricin/TNF-α or ricin/FasL (Figure S7).

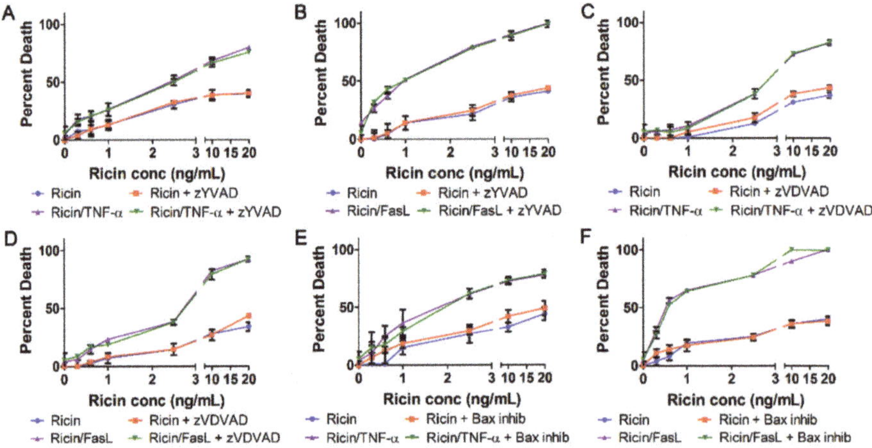

Figure 5. A549 cell death induced by ricin in combination with TNF-α or FasL does not depend on caspase-1, -2, or Bax. A549 lung epithelial cells were treated with ricin alone or in combination with 100 ng/mL TNF-α or FasL in the presence or absence of various pharmacologic inhibitors. Cell death induced by the combination of ricin with TNF-α or FasL was not prevented by inhibition of (**A,B**) caspase-1 with zYVAD-fmk (10 μM), (**C,D**) caspase-2 with zVDVAD-fmk (50 μM), or (**E,F**) Bax with a peptide-based inhibitor (Bax-inhibiting peptide v5, 100 μM). Results are the average of 3 independent experiments. Error bars = standard deviation. Two-way ANOVA, *** $p < 0.001$.

Necroptosis is another major pathway of cell death in addition to apoptosis which may be induced by TNF-α and FasL [23,43]. Thus, we investigated the involvement of RIP1, the initiator kinase of necroptosis [23,43], in cell death by ricin combined with TNF-α or FasL. Cell death induced by ricin combined with either cytokine was not prevented by the RIP1 inhibitor, necrostatin-1s (Figure S8). Collectively, the results of Figure 5 and Figure S7 indicate that ricin/TNF-α and ricin/FasL do not induce one of the classical non-apoptotic cell death pathways. Additionally, the pattern of Mcl1 protein loss does not differ between ricin/TRAIL vs. ricin/TNF-α or ricin/FasL in A549 cells (Figure S9). Thus, we believe the distinct cell death responses to ricin combined with different cytokines is not the result of altered kinetics of protein synthesis inhibition.

2.5. Cell Death Induced by Ricin/TNF-α and Ricin/FasL Depends on Cathepsins

Treatment with zVAD-fmk (pan-caspase inhibitor) prevented cell death induced by ricin/TNF-α and ricin/FasL (Figure 1C,D) yet it does not appear as though caspases are activated during these instances of cell death (Figures 2 and 4, Figure S2 and Figure S5). Importantly, zVAD-fmk may have off-target effects against cathepsins in addition to its specific effects on caspases [44,45]. Therefore, we wondered if the effects of this inhibitor were due to deactivation of cathepsins. Interestingly, zFA-fmk, an inhibitor of executioner caspases-2, -3, -6, and -7 [46] as well as cathepsins B, L and S [47], caused significant inhibition of cell death induced by ricin combined with either TNF-α or FasL (Figure 6A,B). In addition, E64d, an inhibitor of calpains as well as cathepsins B, H, and L [48], blunted A549 cell death by ricin/TNF-α or ricin/FasL (Figure 6C,D). However, calpeptin, an inhibitor of calpains as well as cathepsins L and K [49], had no effect on cell death by ricin/TNF-α or ricin/FasL

(Figure S10). Cathepsin inhibitor 1 (CATI-1) has activity against cathepsins L and S with cathepsin B being its primary target [50]. Inhibition with CATI-1 prevented A549 cell death by ricin combined with TNF-α or FasL (Figure 6E,F). Importantly, CATI-1 had not effect on apoptosis induced by ricin/TRAIL (Figure S11). Cathepsin-dependent cell death is often associated with a dependence on reactive oxygen species (ROS) [51]. We investigated the involvement of ROS in cell death induced by ricin/TNF-α and ricin/FasL using the antioxidant, N-acetylcysteine. Scavenging of ROS by N-acetylcysteine (NAC) resulted in a significant prevention of cell death induced by ricin/TNF-α or ricin/FasL (Figure S12). Collectively, these results indicate that when combined with TNF-α or FasL, ricin induces cell death that depends on cathepsins with a likely role for ROS.

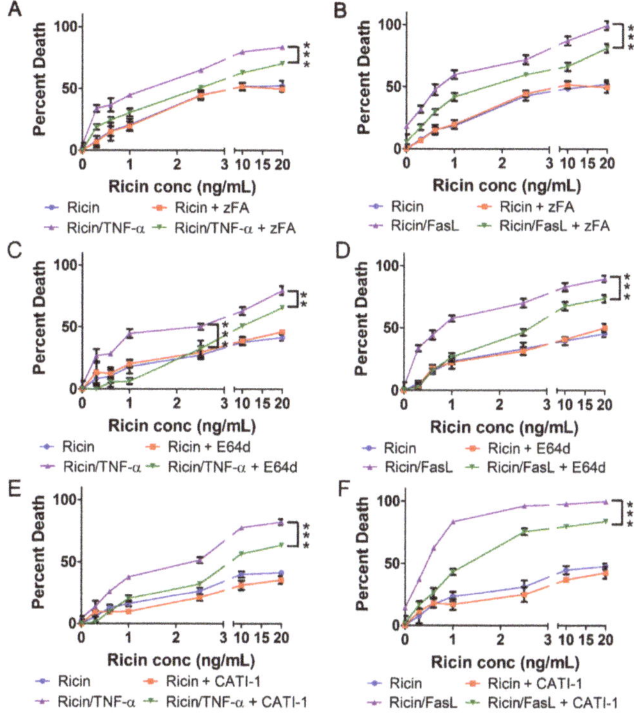

Figure 6. Cell death induced by ricin/TNF-α or ricin/FasL depends on cathepsins. A549 lung epithelial cells were treated with ricin alone or in combination with 100 ng/mL TNF-α or FasL for 24 h at 37 °C followed by measurement of cell death via WST-1 assay. Cell death induced by ricin/TNF-α and ricin/FasL was partially prevented by inhibition of cathepsins with (**A,B**) zFA-fmk (50 μM), (**C,D**) E64d (50 μM), or (**E,F**) cathepsin inhibitor 1 (CATI-1, 20 μM). Results are the average of 3 independent experiments. Error bars = standard deviation. ANOVA, *** $p < 0.001$, ** $p < 0.01$.

The dominant cell line used in this study was A549 human lung epithelial cells, a cell line derived from lung carcinoma. These cells were chosen as they represent a model of human alveolar, type II pneumocytes for drug and toxin metabolism [52,53]. To determine if our results were unique to A549 cells, we conducted experiments with Calu3 human lung epithelial cells. Previously, we showed that ricin/TRAIL induces caspase-dependent apoptosis in Calu3 cells [20]. Here we show that Calu3 cells appear to be insensitive to cell death by ricin/TNF-α or ricin/FasL (Figure S13). Therefore we tested another human cell line, U937 monocytes, as alveolar macrophages are another target cell of ricin upon inhalation. Indeed, ricin/TRAIL induced caspase-dependent apoptosis in U937 cells while ricin/TNF-α and ricin/FasL induced cathepsin-dependent cell death (Figure S14). While the results obtained in

U937 cells supports our findings in A549 cells, these are also a cell line derived from cancerous tissue. Future work should focus on comparing these findings to those obtained in primary lung epithelial cells and macrophages.

3. Discussion

3.1. Distinct Cell Death Modalities

Our results indicate that ricin induces distinct cell death modalities in A549 human lung epithelial cells depending upon the TNF family cytokine with which it is paired. As ricin is the common factor in each of these scenarios, it is tempting to speculate that the cytokines administered with ricin and their downstream signaling pathways are the primary factors in producing these distinct outcomes. However, when combined with CHX, another molecule that inhibits protein synthesis [54], TRAIL, TNF-α, and FasL all induced caspase-dependent apoptosis (Figures 3E and 4G–H). Therefore, there is something unique about the combination of ricin with each of these cytokines which produces different cell death responses. While ricin and CHX both inhibit protein synthesis, they do so by distinct mechanisms: ricin via depurination of rRNA [10,11,13,14] and CHX via occupation of the ribosomal E site [54]. It is possible that these fine mechanistic differences account for the distinct outcomes of cell death depending upon whether TNF family cytokines are paired with ricin or CHX. In addition, there is evidence that the protein synthesis inhibition activity of ricin does not correlate with execution of cell death in other systems [55]. The fact that the cell death caused by ricin does not correlate with protein synthesis inhibition may also provide an explanation for the difference in these cell death outcomes. While TRAIL, TNF-α, and FasL have common upstream signaling events, each pathway eventually diverges [56,57]. Thus, there is precedent for induction of distinct cell death pathways in response to TRAIL, TNF-α, and FasL. The final intersection point in these signaling pathways is likely the death-inducing signaling complex (DISC) [58,59]. We speculate that following DISC formation the signaling pathway induced by ricin/TRAIL diverges from that induced by ricin/TNF-α and ricin/FasL accounting for these distinct cell death outcomes. Future work will include a detailed study on these signaling responses to address this.

Of note is the fact that ricin alone did not appear to induce caspase-dependent apoptosis in A549 cells throughout this work. This is in contrast to in vivo findings as well as results from other human cell types [60,61]. In HeLa cervical cells, MCF7 mammary cells, and U937 monocytes, ricin alone induces caspase-dependent apoptosis [60,61]. The cell death that ricin induces in human lung epithelial cell lines is less certain. The inability of ricin alone to induce apoptosis in A549 cells does not appear to be a unique feature of this cell line as we obtained similar results in Calu3 lung epithelial cells [20]. Therefore, while ricin induces apoptosis in other human cell types, lung epithelial cells may exhibit a different cell death modality in response to this toxin.

3.2. Cathepsin-Dependent Cell Death

We originally hypothesized that the combination of ricin with TRAIL, TNF-α, or FasL would produce caspase-dependent apoptosis. However, this was only the case when ricin was combined with TRAIL (Figures 2A and 3). The fact that the pan-caspase inhibitor, zVAD-fmk, prevented cell death by ricin/TNF-α and ricin/FasL suggested that, like TRAIL, these cytokines also provoke caspase-dependent apoptosis when administered with ricin (Figure 1C,D). However, it is clear that there is no role for caspases in cell death by ricin/TNF-α or ricin/FasL (Figure 4A–F and Figure S5). Moreover, these signaling molecules remain inactivated in response to ricin/TNF-α and ricin/FasL (Figure 2B,C and Figure S2). In addition, we confirmed that alternative caspases (caspases-1 and -2) and downstream apoptosis effectors (Bax) do not play a role in cell death by ricin/TNF-α or ricin/FasL (Figure 5).

Cathepsins are proteases originally discovered to be resident to lysosomes but are present in the cytoplasm and nucleus as well [62]. These molecules may have roles in apoptosis, necroptosis, and pyroptosis [51,62,63]. The results presented in Figure 6 clearly indicate that ricin/TNF-α and ricin/FasL each induce cathepsin-dependent death of A549 human lung epithelial cells. However, we believe that this is a distinct cathepsin-driven cell death pathway and not a component of another cell death pathway (e.g., apoptosis, necroptosis, etc.). Indeed, we ruled out contributions from apoptosis (Figures 2B,C and 4), necroptosis (Figure S2), and pyroptosis (Figure 5A,B) to cell death induced by ricin/TNF-α and ricin/FasL. Cathepsin-driven cell death is often accompanied by ROS activity [51]. Therefore, our findings that N-acetylcysteine (NAC) prevents cell death by ricin/TNF-α or ricin/FasL (Figure S4) are consistent with known mechanisms of cathepsin-dependent cell death, particularly since we have ruled out contributions from apoptosis, necroptosis, and pyroptosis (Figures 4 and 5A,B and Figure S8). It is possible that the cathepsin-dependent cell death induced by ricin/TNF-α and ricin/FasL is a form of lysosome-dependent cell death, which requires the release of cathepsins from lysosomes via permeabilization [51,63]. However, cathepsins are known to be resident to the cytoplasm as well [62], making lysosomal involvement in these pathways unclear. If lysosomal permeabilization is the basis for cathepsin involvement in these pathways, it is likely to be ROS-mediated, particularly in light of the results from Figure S12. The other major stimulus for lysosomal permeabilization is Bak/Bax [51], whose involvement we have ruled out (Figure 5E,F).

Interestingly, in Figure S11 NAC did not prevent cell death induced by ricin alone. This is in contrast to other reports that ROS are involved in cell death induced by ricin [64,65]. Data on the involvement of ROS in ricin-induced death of human lung epithelial cells are lacking, however. As noted earlier, ricin alone clearly induces caspase-dependent apoptosis in human monocytes, mammary cells, and cervical cells [60,61]. However, in this work as well as our prior work we have shown that ricin alone does not induce caspase-dependent apoptosis of human lung epithelial cells lines A549 (Figures 2–4) and Calu3 [20]. We hypothesize that NAC did not affect cell death by ricin alone in A549 cells as these cells undergo a cell death response to ricin which differs from the caspase-dependent apoptosis that ricin induces in other human cell types. This also agrees with the accepted concept that ROS have a key role in apoptosis [66].

3.3. Therapeutic Implications

Enhancement of ricin-induced cell death by TNF family cytokines presents new potential therapeutic targets against ricin toxicity. Targeting these cell death ligands directly with neutralizing antibodies represents a viable therapeutic option as does targeting ricin with neutralizing antibodies. We explored this previously at the cellular level [20]. Our identification of activated cell death pathways in response to ricin combined with TNF family cytokines reveals further potential therapeutic targets (i.e., specific components of each cell death pathway). However, the finding that different cytokines induce distinct cell death pathways in combination with ricin suggests that a combinatorial approach may be best in the context of ricin toxicity. Rather than targeting apoptosis, which would limit ricin/TRAIL-induced cell death only, both caspase-dependent apoptosis and cathepsin-dependent cell death should be targeted. In addition, we cannot rule out contributions from other cytokines and cell death pathways to ricin-induced cell death in the in vivo setting. This could result in the targeting of multiple cell death pathways at various steps to limit ricin-induced toxicity. Another potential therapeutic implication of this work may extend to the targeting of tumors, as we have utilized cell lines derived from cancerous tissue throughout this research. Recent research on cancer therapeutics has partly focused on transfection of cytotoxic genes into tumor cells [67]. Ricin may serve as a good candidate for such targeted cancer therapy, particularly if combined with administration of TNF family cytokines.

4. Materials and Methods

4.1. Reagents and Inhibitors

Ricin toxin (*Ricinus communis* agglutinin II) was purchased from Vector Laboratories and used at the concentrations noted. Ricin was dialyzed in 1× PBS using 10K MW-cutoff Slide-A-Lyzer dialysis cassettes (Pierce, Rockford, IL, USA) prior to experimentation. Cycloheximide (Sigma, St. Louis, MO, USA) was used at a concentration of 500 ng/mL unless noted otherwise. Recombinant human TRAIL (Peprotech, Rocky Hill, NJ, USA), TNF-α (Shenandoah Biotechnology, Warwick, PA, USA), and FasL (Super Fas ligand, Enzo Life Sciences, Farmingdale, NY, USA) were used at a concentration of 100 ng/mL unless noted otherwise. All caspase inhibitors are irreversible and were purchased from ApexBio (Houston, TX, USA). Pan-caspase inhibitor zVAD-fmk, executioner caspase and cathepsin inhibitor zFA-fmk, and caspase-2 inhibitor zVDVAD were each used at a concentration of 50 μM. Caspase-3/7 inhibitor zDEVD, caspase-8 inhibitor zIETD, and caspase-6 inhibitor zVEID were each used at a concentration of 30 μM. Caspase-1 inhibitor zYVAD-fmk and caspase-9 inhibitor zLEHD-fmk were each used at a concentration of 10 μM. Necrostatin-1s (EMD Millipore, Burlington, MA, USA) was used at a concentration of 50 μM. Bax-inhibiting peptide v5 (EMD Millipore) was used at a concentration of 100 μM. E64d was used at a concentration of 10 μM. Cathepsin inhibitor 1 (CATI-1, ApexBio) was used at a concentration of 20 μM. N-acetylcysteine (Sigma) was used at a concentration of 10 mM.

4.2. Cell Culture

A549 lung epithelial cells (CCL-185, ATCC, Manassas, VA, USA) were cultured in Ham's F-12K medium (Thermo Fisher) with 10% FBS as recommended by the ATCC. Cells were grown in a humidified incubator with 5% CO_2 at 37 °C. Cells were cultured in 75-cm^2 cell culture flasks and subcultured when they reached ~80% confluence at a 1:5 dilution. Cells were lifted using TrypLE Express (Thermo Fisher, Waltham, MA, USA) at 37 °C. Cells were maintained at a maximum of 10 passages for the experiments in this work.

4.3. Cell Death Assays

A549 lung epithelial cells were seeded at 1200 cells/well in 96-well plates (Celltreat, Pepperell, MA, USA) and allowed to culture for 24 h in Ham's F-12K medium (Thermo Fisher) with 10% FBS (VWR, Radnor, PA, USA) at 37 °C and 5% CO_2. Following this, cells were washed and the following was added to wells in Ham's F-12K medium: (1) ricin, (2) 100 ng/mL cell death ligand/cytokine (TRAIL, TNF-α, or FasL), or (3) ricin combined with 100 ng/mL cell death ligand/cytokine. Cells in negative control wells were treated with media alone. In some experiments (Figure 1B), CHX was used in place of ricin. In experiments where inhibitors were used, they were added to cells 1 h before the addition of ricin and cell death ligands/cytokines. Appropriate vehicle controls were used for each inhibitor. After the addition of ricin and cell death ligands/cytokines, cells were incubated for 24 h at 37 °C and 5% CO_2. Cell death was then measured using the WST-1 assay (Takara, Kusatsu, Shiga Prefecture, Japan) according to the manufacturer's instructions. Absorbance was measured using an Eppendorf 2200 plate reader at a wavelength of 450 nm and a reference wavelength of 600 nm. Using WST-1 absorbance (abs), percent viability was calculated as follows: (abs cell death stimulus + ricin)/(abs neg) × 100.

4.4. Immunoblots

A total of 6×10^6 A549 cells were seeded in a 58 cm^2 Petri dish (Celltreat) per condition and allowed to grow for 24 h at 37 °C and 5% CO_2. A549 cells were then treated with (1) 1 ng/mL ricin, (2) 100 ng/mL cell death ligand/cytokine (TRAIL, TNF-α, or FasL), or 3.) 1 ng/mL ricin combined with 100 ng/mL cell death ligand/cytokine for 4 h at 37 °C and 5% CO_2. Following this, A549 cells were lysed using 1% triton-X-100 (Sigma) with 1× Halt protease inhibitor (Thermo Fisher) in 1× PBS on ice for 30 min. Cells were then sonicated on ice 3× (20 sec pulses) at an output of 10%. Cell lysates were centrifuged at 14000 rpm at 4 °C to remove nuclear material. A549 cell lysates were run on SDS-PAGE and transferred to a PVDF membrane and blocked in 1× TBS with 0.1% tween-20 and 5% milk for 30 min at room temperature. The blots were then incubated with diluted primary antibody in 1× TBS with 0.1% tween-20 and 5% milk overnight at 4 °C. All primary antibodies were obtained from Cell Signaling Technology (Danvers, MA, USA), unless otherwise indicated. Primary antibodies were used at the following dilutions: anti-human caspase-3 (1:1000), anti-human/mouse caspase-6 (1:1000), anti-human/mouse caspase-7 (1:1000), anti-human caspase-8 (1:1000), anti-human caspase-9 (1:1000), and anti-human GAPDH (1:10,000). After washing with 1× TBS with 0.1% tween-20 and 5% milk, the blots were incubated with secondary HRP-conjugate antibodies (1:5000) for 1 h at room temperature. Blots were developed by chemiluminescence and read in a Bio-Rad ChemiDoc XRS+.

4.5. Statistical Analyses

Statistical analyses were carried out using GraphPad Prism 7. All cell death assays were subject to statistical analysis by two-way ANOVA and Bonferroni posttest. All cell death assays are the results of 3 independent experiments. Immunoblots presented are representative of 3 independent experiments.

Supplementary Materials: The following are available online at http://www.mdpi.com/2072-6651/11/8/450/s1, Figure S1: The combination of ricin/TRAIL induces caspase activation, Figure S2: Ricin/TNF-α and ricin/FasL do not induce caspase activation in A549 cells after 8 h, Figure S3: The apoptotic stimuli CHX/TRAIL, CHX/TNF-α, and CHX/FasL induce caspase cleavage, Figure S4: Ricin/TRAIL induces caspase cleavage that is prevented by pharmacologic inhibitors, Figure S5: A549 cell death by ricin/TNF-α and ricin/FasL does not depend on caspases-6 or -9, Figure S6: The apoptotic stimuli CHX/TNF-α and CHX/FasL induce Bax-dependent cell death, Figure S7: Cell death induced by ricin in combination with TNF-α or FasL does not depend on RIP1 kinase, Figure S8: Mcl1 protein loss does not differ between ricin/TRAIL, ricin/TNF-α, and ricin/FasL, Figure S9: Cell death induced by ricin/TNF-α and ricin/FasL does not depend on calpains or cathepsins L and K, Figure S10: The cathepsin inhibitor, CATI-1, has no effect on A549 cell death by ricin/TRAIL, Figure S11: Cell death induced by ricin/TNF-α or ricin/FasL is inhibited by N-acetylcysteine, Figure S12: Calu3 human lung epithelial cells are insensitive to cell death by ricin/TNF- or ricin/FasL, Figure S13: Calu3 human lung epithelial cells are insensitive to cell death by ricin/TNF-α or ricin/FasL,. Figure S14: Ricin/TRAIL induces caspase-dependent apoptosis while ricin/TNF-α and ricin/FasL induce cathepsin-dependent cell death in U937 human monocytes.

Author Contributions: Conceptualization, T.J.L. and N.J.M.; methodology, T.J.L. and W.D.M.; formal analysis, T.J.L., A.L.H., and C.G.K.; investigation, A.L.H., C.G.K., and C.A.P.; data curation, T.J.L.; writing—original draft preparation, T.J.L.; writing—review and editing, T.J.L., A.L.H., W.D.M., and N.J.M.; supervision, T.J.L. and W.D.M.; project administration, T.J.L.; funding acquisition, T.J.L, N.J.M.

Funding: This research was funded by the National Heart, Lung, and Blood Institute (NHLBI) of the National Institutes of Health (NIH), grant number NIH R15-HL135675-01, awarded to T.J.L and the National Institutes of Allergy and Infectious Diseases (NIAID) of the NIH, contract number HHSN272201400021C, awarded to N.J.M. The content is solely the responsibility of the authors and does not necessarily represent the official views of the NIH. The funders had no role in study design, data collection and analysis, decision to publish, or preparation of the manuscript.

Conflicts of Interest: The authors declare no conflict of interest.

References

1. Bradberry, S.M.; Dickers, K.J.; Rice, P.; Griffiths, G.D.; Vale, J.A. Ricin poisoning. *Toxicol. Rev.* **2003**, *22*, 65–70. [CrossRef] [PubMed]
2. Audi, J.; Belson, M.; Patel, M.; Schier, J.; Osterloh, J. Ricin poisoning a comprehensive review. *J. Am. Med. Assoc.* **2005**, *294*, 2342–2351. [CrossRef] [PubMed]
3. Lopez Nunez, O.F.; Pizon, A.F.; Tamama, K. Ricin Poisoning after Oral Ingestion of Castor Beans: A Case Report and Review of the Literature and Laboratory Testing. *J. Emerg. Med.* **2017**, *53*, e67–e71. [CrossRef] [PubMed]
4. Anderson, P.D. Bioterrorism: Toxins as weapons. *J. Pharm. Pract.* **2012**, *25*, 121–129. [CrossRef] [PubMed]
5. Gopalakrishnakone, P.; Balali-Mood, M.; Llewellyn, L.; Singh, B.R. *Biological Toxins and Bioterrorism*; Springer: Dordrecht, The Netherlands, 2015; ISBN 9789400758698.
6. Falach, R.; Sapoznikov, A.; Gal, Y.; Israeli, O.; Leitner, M.; Seliger, N.; Ehrlich, S.; Kronman, C.; Sabo, T. Quantitative profiling of the in vivo enzymatic activity of ricin reveals disparate depurination of different pulmonary cell types. *Toxicol. Lett.* **2016**, *258*, 11–19. [CrossRef] [PubMed]
7. Lord, J.M.; Roberts, L.M.; Robertus, J.D. Ricin: structure, mode of action, and some current applications. *FASEB J.* **2018**, *8*, 201–208. [CrossRef]
8. Taubenschmid, J.; Stadlmann, J.; Jost, M.; Klokk, T.I.; Rillahan, C.D.; Leibbrandt, A.; Mechtler, K.; Paulson, J.C.; Jude, J.; Zuber, J.; et al. A vital sugar code for ricin toxicity. *Cell Res.* **2017**, *27*, 1351–1364. [CrossRef] [PubMed]
9. Spooner, R.A.; Michael Lord, J. Ricin trafficking in cells. *Toxins (Basel)* **2015**, *7*, 49–65. [CrossRef] [PubMed]

10. Spooner, R.A.; Watson, P.D.; Marsden, C.J.; Smith, D.C.; Moore, K.A.H.; Cook, J.P.; Lord, J.M.; Roberts, L.M. Protein disulphide-isomerase reduces ricin to its A and B chains in the endoplasmic reticulum. *Biochem. J.* **2004**, *383*, 285–293. [CrossRef]

11. Endo, Y.; Tsurugi, K. The RNA N-glycosidase activity of ricin A-chain. *Nucleic Acids Symp. Ser.* **1988**, *19*, 139–142.

12. RNA N-glycosidase activity of ricin A-chain. Mechanism of action of the toxic lectin ricin on eukaryotic ribosomes. *J. Biol. Chem.* **1987**, *262*, 8128–8130.

13. Sperti, S.; Montanaro, L.; Mattioli, A.; Stirpe, F. Inhibition by ricin of protein synthesis in vitro: 60S ribosomal subunit as the target of the toxin (Short Communication). *Biochem. J.* **1973**, *136*, 813–815. [CrossRef] [PubMed]

14. Shi, X.; Khade, P.K.; Sanbonmatsu, K.Y.; Joseph, S. Functional role of the sarcin-ricin loop of the 23s rRNA in the elongation cycle of protein synthesis. *J. Mol. Biol.* **2012**, *419*, 125–138. [CrossRef]

15. Griffiths, G.D.; Phillips, G.J.; Holley, J. Inhalation toxicology of ricin preparations: Animal models, prophylactic and therapeutic approaches to protection. *Inhal. Toxicol.* **2007**, *19*, 873–887. [CrossRef] [PubMed]

16. Wilhelmsen, C.L.; Pitt, M.L.M. Lesions of acute inhaled lethal ricin intoxication in rhesus monkeys. *Vet. Pathol.* **1996**, *33*, 296–302. [CrossRef] [PubMed]

17. Matthay, M.A.; Zemans, R.L.; Zimmerman, G.A.; Arabi, Y.M.; Beitler, J.R.; Mercat, A.; Herridge, M.; Randolph, A.G.; Calfee, C.S. Acute respiratory distress syndrome. *Nat. Rev. Dis. Prim.* **2019**, *5*, 18. [CrossRef] [PubMed]

18. Pincus, S.H.; Bhaskaran, M.; Brey, R.N.; Didier, P.J.; Doyle-Meyers, L.A.; Roy, C.J. Clinical and pathological findings associated with aerosol exposure of macaques to ricin toxin. *Toxins (Basel)* **2015**, *7*, 2121–2133. [CrossRef]

19. Gal, Y.; Mazor, O.; Falach, R.; Sapoznikov, A.; Kronman, C.; Sabo, T. Treatments for pulmonary ricin intoxication: Current aspects and future prospects. *Toxins (Basel)* **2017**, *9*, 311. [CrossRef]

20. Rong, Y.; Westfall, J.; Ehrbar, D.; LaRocca, T.; Mantis, N.J. TRAIL (CD253) Sensitizes Human Airway Epithelial Cells to Toxin-Induced Cell Death. *mSphere* **2018**, *3*, e00399-18. [CrossRef]

21. Taylor, R.C.; Cullen, S.P.; Martin, S.J. Apoptosis: Controlled demolition at the cellular level. *Nat. Rev. Mol. Cell Biol.* **2008**, *9*, 231–241. [CrossRef]

22. Pasparakis, M.; Vandenabeele, P. Necroptosis and its role in inflammation. *Nature* **2015**, *517*, 311–320. [CrossRef]

23. Vandenabeele, P.; Galluzzi, L.; Vanden Berghe, T.; Kroemer, G. Molecular mechanisms of necroptosis: an ordered cellular explosion. *Nat. Rev. Mol. Cell Biol.* **2010**, *11*, 700–714. [CrossRef]

24. Korcheva, V.; Wong, J.; Lindauer, M.; Jacoby, D.B.; Iordanov, M.S.; Magun, B. Role of apoptotic signaling pathways in regulation of inflammatory responses to ricin in primary murine macrophages. *Mol. Immunol.* **2007**, *44*, 2761–2771. [CrossRef]

25. Wong, J.; Korcheva, V.; Jacoby, D.B.; Magun, B. Intrapulmonary delivery of ricin at high dosage triggers a systemic inflammatory response and glomerular damage. *Am. J. Pathol.* **2007**, *170*, 1497–1510. [CrossRef]

26. DaSilva, L.; Cote, D.; Roy, C.; Martinez, M.; Duniho, S.; Pitt, M.L.M.; Downey, T.; Dertzbaugh, M. Pulmonary gene expression profiling of inhaled ricin. *Toxicon* **2003**, *41*, 813–822. [CrossRef]

27. David, J.; Wilkinson, L.J.; Griffiths, G.D. Inflammatory gene expression in response to sub-lethal ricin exposure in Balb/c mice. *Toxicology* **2009**, *264*, 119–130. [CrossRef]

28. Lindauer, M.; Wong, J.; Magun, B. Ricin toxin activates the NALP3 inflammasome. *Toxins (Basel)* **2010**, *2*, 1500–1514. [CrossRef]

29. Wong, J.; Magun, B.E.; Wood, L.J. Lung inflammation caused by inhaled toxicants: A review. *Int. J. COPD* **2016**, *11*, 1391–1401. [CrossRef]

30. Knight, M.J.; Riffkin, C.D.; Muscat, A.M.; Ashley, D.M.; Hawkins, C.J. Analysis of FasL and trail induced apoptosis pathways in glioma cells. *Oncogene* **2001**, *20*, 5789–5798. [CrossRef]

31. Laster, S.M.; Wood, J.G.; Gooding, L.R. Tumor necrosis factor can induce both apoptic and necrotic forms of cell lysis. *J. Immunol.* **1988**, *141*, 2629–2634.

32. Babu, D.J.; Soenen, S.; Raemdonck, K.; Leclercq, G.; De Backer, O.; Motterlini, R.A.; Lefebvre, R. TNF-α/Cycloheximide-Induced Oxidative Stress and Apoptosis in Murine Intestinal Epithelial MODE-K Cells. *Curr. Pharm. Des.* **2012**, *18*, 4414–4425. [CrossRef]

33. Jin, S.; Ray, R.M.; Johnson, L.R. TNF-α/cycloheximide-induced apoptosis in intestinal epithelial cells requires Rac1-regulated reactive oxygen species. *Am. J. Physiol. Liver Physiol.* **2008**, *294*, G928–G937. [CrossRef]

34. Vercammen, D.; Vandenabeele, P.; Beyaert, R.; Declercq, W.; Fiers, W. Tumour necrosis factor-induced necrosis versus anti-Fas-induced apoptosis in L929 cells. *Cytokine* **1997**, *9*, 801–808. [CrossRef]

35. Lin, Y.; Choksi, S.; Shen, H.M.; Yang, Q.F.; Hur, G.M.; Kim, Y.S.; Tran, J.H.; Nedospasov, S.A.; Liu, Z.G. Tumor Necrosis Factor-induced Nonapoptotic Cell Death Requires Receptor-interacting Protein-mediated Cellular Reactive Oxygen Species Accumulation. *J. Biol. Chem.* **2004**, *279*, 10822–10828. [CrossRef]

36. LaRocca, T.J.; Sosunov, S.A.; Shakerley, N.L.; Ten, V.S.; Ratner, A.J. Hyperglycemic conditions prime cells for RIP1-dependent necroptosis. *J. Biol. Chem.* **2016**, *291*, 13753–13761. [CrossRef]

37. McCaig, W.D.; Patel, P.S.; Sosunov, S.A.; Shakerley, N.L.; Smiraglia, T.A.; Craft, M.M.; Walker, K.M.; Deragon, M.A.; Ten, V.S.; LaRocca, T.J. Hyperglycemia potentiates a shift from apoptosis to RIP1-dependent necroptosis. *Cell Death Discov.* **2018**, *4*, 55. [CrossRef]

38. Mori, T.; Doi, R.; Toyoda, E.; Koizumi, M.; Ito, D.; Kami, K.; Kida, A.; Masui, T.; Kawaguchi, Y.; Fujimoto, K. Regulation of the resistance to TRAIL-induced apoptosis as a new strategy for pancreatic cancer. *Surgery* **2005**, *138*, 71–77. [CrossRef]

39. Hellwig, C.T.; Kohler, B.F.; Lehtivarjo, A.K.; Dussmann, H.; Courtney, M.J.; Prehn, J.H.M.; Rehm, M. Real time analysis of tumor necrosis factor-related apoptosis-inducing ligand/cycloheximide-induced caspase activities during apoptosis initiation. *J. Biol. Chem.* **2008**, *283*, 21676–21685. [CrossRef]

40. Shi, J.; Gao, W.; Shao, F. Pyroptosis: Gasdermin-Mediated Programmed Necrotic Cell Death. *Trends Biochem. Sci.* **2017**, *42*, 245–254. [CrossRef]

41. Kovacs, S.B.; Miao, E.A. Gasdermins: Effectors of Pyroptosis. *Trends Cell Biol.* **2017**, *27*, 673–684. [CrossRef]

42. Bouchier-Hayes, L. The role of caspase-2 in stress-induced apoptosis. *J. Cell. Mol. Med.* **2010**, *14*, 1212–1224. [CrossRef]

43. Linkermann, A.; Green, D.R. Necroptosis. *N. Engl. J. Med.* **2014**, *370*, 455–465. [CrossRef]

44. Lawrence, C.P.; Kadioglu, A.; Yang, A.-L.; Coward, W.R.; Chow, S.C. The Cathepsin B Inhibitor, z-FA-FMK, Inhibits Human T Cell Proliferation In Vitro and Modulates Host Response to Pneumococcal Infection In Vivo. *J. Immunol.* **2006**, *177*, 3827–3836. [CrossRef]

45. Rozman-Pungerčar, J.; Kopitar-Jerala, N.; Bogyo, M.; Turk, D.; Vasiljeva, O.; Stefe, I.; Vandenabeele, P.; Brömme, D.; Puizdar, V.; Fonović, M.; et al. Inhibition of papain-like cysteine proteases and legumain by caspase-specific inhibitors: When reaction mechanism is more important than specificity. *Cell Death Differ.* **2003**, *10*, 881–888. [CrossRef]

46. Lopez-Hernandez, F.J.; Ortiz, M.A.; Bayon, Y.; Piedrafita, F.J. Z-FA-fmk inhibits effector caspases but not initiator caspases 8 and 10, and demonstrates that novel anticancer retinoid-related molecules induce apoptosis via the intrinsic pathway. *Mol. Cancer Ther.* **2003**, *2*, 255–263.

47. Ahmed, N.K.; Martin, L.A.; Watts, L.M.; Palmer, J.; Thornburg, L.; Prior, J.; Esser, R.E. Peptidyl fluoromethyl ketones as inhibitors of cathepsin B. Implication for treatment of rheumatoid arthritis. *Biochem. Pharmacol.* **1992**, *44*, 1201–1207. [CrossRef]

48. Matsumoto, K.; Mizoue, K.; Kitamura, K.; Tse, W.C.; Huber, C.P.; Ishida, T. Structural basis of inhibition of cysteine proteases by E-64 and its derivatives. *Biopolym. Pept. Sci. Sect.* **1999**, *51*, 99–107. [CrossRef]

49. Siklos, M.; BenAissa, M.; Thatcher, G.R.J. Cysteine proteases as therapeutic targets: Does selectivity matter? A systematic review of calpain and cathepsin inhibitors. *Acta Pharm. Sin. B* **2015**, *5*, 506–519. [CrossRef]

50. Demuth, H.U.; Schierhorn, A.; Bryan, P.; Höfke, R.; Kirschke, H.; Brömme, D. N-peptidyl, O-acyl hydroxamates: Comparison of the selective inhibition of serine and cysteine proteinases. *Biochim. Biophys. Acta—Protein Struct. Mol. Enzymol.* **1996**, *1295*, 179–186. [CrossRef]

51. Wang, F.; Gómez-Sintes, R.; Boya, P. Lysosomal membrane permeabilization and cell death. *Traffic* **2018**, *19*, 918–931. [CrossRef]

52. Cooper, J.R.; Abdullatif, M.B.; Burnett, E.C.; Kempsell, K.E.; Conforti, F.; Tolley, H.; Collins, J.E.; Davies, D.E. Long term culture of the a549 cancer cell line promotes multilamellar body formation and differentiation towards an alveolar type II Pneumocyte phenotype. *PLoS ONE* **2016**, *11*, e0164438. [CrossRef]

53. Foster, K.A.; Oster, C.G.; Mayer, M.M.; Avery, M.L.; Audus, K.L. Characterization of the A549 cell line as a type II pulmonary epithelial cell model for drug metabolism. *Exp. Cell Res.* **1998**, *243*, 359–366. [CrossRef]

54. Schneider-Poetsch, T.; Ju, J.; Eyler, D.E.; Dang, Y.; Bhat, S.; Merrick, W.C.; Green, R.; Shen, B.; Liu, J.O. Inhibition of eukaryotic translation elongation by cycloheximide and lactimidomycin. *Nat. Chem. Biol.* **2010**, *6*, 209–217. [CrossRef]

55. Li, X.P.; Baricevic, M.; Saidasan, H.; Tumer, N.E. Ribosome depurination is not sufficient for ricin-mediated cell death in Saccharomyces cerevisiae. *Infect. Immun.* **2007**, *75*, 417–428. [CrossRef]

56. Jin, Z.; El-Deiry, W.S. Distinct Signaling Pathways in TRAIL- versus Tumor Necrosis Factor-Induced Apoptosis. *Mol. Cell. Biol.* **2006**, *26*, 8136–8148. [CrossRef]

57. Bang, S.; Jeong, E.J.; Kim, I.K.; Jung, Y.K.; Kim, K.S. Fas- and tumor necrosis factor-mediated apoptosis uses the same binding surface of FADD to trigger signal transduction: A typical model for convergent signal transduction. *J. Biol. Chem.* **2000**, *275*, 36217–36222. [CrossRef]

58. Guicciardi, M.E.; Gores, G.J. Life and death by death receptors. *FASEB J.* **2009**, *23*, 1625–1637. [CrossRef]

59. Lavrik, I. Death receptor signaling. *J. Cell Sci.* **2005**, *118*, 265–267. [CrossRef]

60. Sikriwal, D.; Batra, J.K. Ribosome inactivating proteins and apoptosis. *Plant Cell Monogr.* **2010**, *18*, 167–189.

61. Tesh, V.L. The induction of apoptosis by Shiga toxins and ricin. *Curr. Top. Microbiol. Immunol.* **2012**, *357*, 137–178.

62. Turk, V.; Stoka, V.; Vasiljeva, O.; Renko, M.; Sun, T.; Turk, B.; Turk, D. Cysteine cathepsins: From structure, function and regulation to new frontiers. *Biochim. Biophys. Acta Proteins Proteomics* **2012**, *1824*, 68–88. [CrossRef]

63. Boya, P.; Kroemer, G. Lysosomal membrane permeabilization in cell death. *Oncogene* **2008**, *27*, 6434–6451. [CrossRef]

64. Suntres, Z.E.; Stone, W.L.; Smith, M.G. Ricin—Induced Toxicity: The Role of Oxidative Stress. *J. Med. CBR Def.* **2005**, *3*, 1–21.

65. Oda, T.; Iwaoka, J.; Komatsu, N.; Tsuyoshi, M. Involvement of N-acetylcysteine-sensitive pathways in ricin-induced apoptotic cell death in U937 cells. *Biosci. Biotechnol. Biochem.* **1999**, *63*, 341–348. [CrossRef]

66. Circu, M.L.; Aw, T.Y. Reactive oxygen species, cellular redox systems, and apoptosis. *Free Radic. Biol. Med.* **2010**, *48*, 749–762. [CrossRef]

67. Ling, C. Cytotoxic genes from traditional Chinese medicine inhibit tumor growth both in vitro and in vivo. *J. Integr. Med.* **2014**, *12*, 483–494.

Article

Generation of Highly Efficient Equine-Derived Antibodies for Post-Exposure Treatment of Ricin Intoxications by Vaccination with Monomerized Ricin

Reut Falach [1], Anita Sapoznikov [1], Ron Alcalay [1], Moshe Aftalion [1], Sharon Ehrlich [1], Arik Makovitzki [2], Avi Agami [2], Avishai Mimran [2], Amir Rosner [3], Tamar Sabo [1], Chanoch Kronman [1] and Yoav Gal [1,*]

[1] Department of Biochemistry and Molecular Genetics, Israel Institute for Biological Research, Ness-Ziona 76100, Israel; reutf@iibr.gov.il (R.F.); anitas@iibr.gov.il (A.S.); rona@iibr.gov.il (R.A.); moshea@iibr.gov.il (M.A.); sharone@iibr.gov.il (S.E.); tamars@iibr.gov.il (T.S.); chanochk@iibr.gov.il (C.K.)
[2] Department of Biotechnology, Israel Institute for Biological Research, Ness-Ziona 76100, Israel; arikm@iibr.gov.il (A.M.); avia@iibr.gov.il (A.A.); avishaim@iibr.gov.il (A.M.)
[3] Veterinary Center for Preclinical Research, Israel Institute for Biological Research, Ness-Ziona 76100, Israel; amirr@iibr.gov.il
* Correspondence: yoavg@iibr.gov.il; Tel.: +972-8-9381479

Received: 25 October 2018; Accepted: 8 November 2018; Published: 12 November 2018

Abstract: Ricin, a highly lethal toxin derived from the seeds of *Ricinus communis* (castor beans) is considered a potential biological threat agent due to its high availability, ease of production, and to the lack of any approved medical countermeasure against ricin exposures. To date, the use of neutralizing antibodies is the most promising post-exposure treatment for ricin intoxication. The aim of this work was to generate anti-ricin antitoxin that confers high level post-exposure protection against ricin challenge. Due to safety issues regarding the usage of ricin holotoxin as an antigen, we generated an inactivated toxin that would reduce health risks for both the immunizer and the immunized animal. To this end, a monomerized ricin antigen was constructed by reducing highly purified ricin to its monomeric constituents. Preliminary immunizing experiments in rabbits indicated that this monomerized antigen is as effective as the native toxin in terms of neutralizing antibody elicitation and protection of mice against lethal ricin challenges. Characterization of the monomerized antigen demonstrated that the irreversibly detached A and B subunits retain catalytic and lectin activity, respectively, implying that the monomerization process did not significantly affect their overall structure. Toxicity studies revealed that the monomerized ricin displayed a 250-fold decreased activity in a cell culture-based functionality test, while clinical signs were undetectable in mice injected with this antigen. Immunization of a horse with the monomerized toxin was highly effective in elicitation of high titers of neutralizing antibodies. Due to the increased potential of IgG-derived adverse events, anti-ricin F(ab′)₂ antitoxin was produced. The F(ab′)₂-based antitoxin conferred high protection to intranasally ricin-intoxicated mice; ~60% and ~34% survival, when administered 24 and 48 h post exposure to a lethal dose, respectively. In line with the enhanced protection, anti-inflammatory and anti-edematous effects were measured in the antitoxin treated mice, in comparison to mice that were intoxicated but not treated. Accordingly, this anti-ricin preparation is an excellent candidate for post exposure treatment of ricin intoxications.

Keywords: ricin; vaccine; antitoxin; RTA; RTB; reduction; alkylation

Key Contribution: In this study, we demonstrate that treatment of mice at clinically relevant time points with anti-ricin F(ab′)₂-based antitoxin derived from a horse immunized with neutralized ricin, conferred exceptionally high protection against a lethal intranasal ricin challenge.

---

1. Introduction

Ricin, a type II ribosome inactivating protein (RIP) derived from the seeds of *Ricinus communis* (castor beans), consists of two polypeptide subunits (A and B) linked by a disulfide bond. The B subunit (RTB) is a lectin, which binds to galactose residues on the cell surface, enabling toxin internalization into the cells. The catalytically active A subunit (RTA) is translocated into the cytoplasm, where it depurinates a conserved adenine residue within the 28S ribosomal RNA of the 60S subunit, leading to irreversible inhibition of protein synthesis and ultimately to cell death [1]. Classified as a Category B agent by the U.S. Centers for Disease Control and Prevention (CDC), ricin is considered a potential bioterror agent mainly due to its high availability and ease of preparation [2]. Ricin toxicity depends on the route of exposure, inhalational and parenteral being highly fatal. Although prophylactic anti-ricin vaccines are being developed [3], the only post-exposure measure found effective against ricin intoxications in pre-clinical settings, is passive immunization with anti-ricin neutralizing antibodies [4–7]. Anti-ricin antibodies may be elicited following vaccination of various animal species, including mice [8], rabbits [9], monkeys [10], horses [11] and sheep [12]. A variety of ricin immunogens were employed to elicit neutralizing antibody responses against ricin. A toxoid-based vaccine (formaldehyde-inactivated ricin) was shown to induce high titers of protective antibodies both in rabbits [4] and in sheep [12]. Anti-ricin preparations were reported to elicit potent toxin neutralization in vitro and in vivo following horse immunization with an RTA/RTB chain construct, in which the native inter-chain linking domain has been replaced by a non-cleavable linker [11]. Vaccination of animals using the native ricin toxin emulsified in adjuvant was also effective in the elicitation of high titers of neutralizing antibodies [13].

As part of ongoing efforts to develop novel yet safe vaccination strategies that will induce high titers of ricin neutralizing antibodies, we established a method for ricin subunit-based immunization following irreversible monomerization of the toxin to its RTA and RTB constituents. The antigen was prepared by treating the toxin with a reducing agent to sever the inter-subunit covalent bond, and then alkylating the monomeric toxin subunits to prevent their re-dimerization, thereby generating a stable monomerized ricin preparation for animal immunization with substantially reduced toxicity.

In the present study, the efficacy potential of the monomerized ricin vaccine was demonstrated in rabbits, after which the antigen, which was produced at large amount, was thoroughly characterized. The antigen, which was used for vaccinating the horse, was not only safe, but also elicited high titers of highly potent neutralizing antibodies against ricin. Passive immunization with the F(ab')$_2$-based antitoxin conferred high protection against a lethal intranasal ricin challenge at clinically relevant treatment time points following intoxication, and displayed significant anti-inflammatory and anti-edematous effects.

2. Results

2.1. Elicitation of Anti-Ricin Antibodies Following Rabbit Immunization with Monomerized Ricin

To determine whether the deactivation of ricin by monomerization, affects its ability to elicit neutralizing antibodies, we compared the anti-ricin antibody titers elicited by immunization either with the monomeric antigen or with native holotoxin. To this end, rabbits were immunized at 4-week intervals, with increasing antigen doses reaching up to 100 µg antigen/rabbit, and hyperimmune sera collected at 16 weeks after the first immunization dose were characterized. As seen (Table 1) high ELISA- and neutralizing-antibody titers were reached following immunization with both native or monomerized ricin (2.56×10^5 vs. 4.8×10^5 ELISA units; 7.68×10^4 vs. 9.6×10^4 neutralizing units, respectively). In line with these findings, hyperimmune sera collected from rabbits immunized with either monomerized or native ricin, conferred similarly high level protection in vivo; when mice were treated intramuscularly (i.m.) with the rabbit antisera 24 h prior to i.m. ricin challenge, the doses which conferred 50% protection (PD$_{50}$) were 0.8 and 1.1 µL/mouse, respectively. These experiments suggest that immunization of a horse with monomerized ricin vaccine would be as effective as immunization with native toxin in elicitation of neutralizing anti-ricin antibodies, while the neutralized monomer-based antigen would possess much greater safety margins.

Table 1. Anti-ricin antibody titers and protective doses of anti-ricin antisera elicited in immunized rabbits.

Antigen	ELISA-Antibody Titer [a]	Neutralizing-Antibody Titer [a]	PD_{50} (μL) [b]
Native Toxin	256,000 ± 41,300	76,800 ± 14,300	1.1 ± 0.31
Monomerized Toxin	480,000 ± 71,500	96,000 ± 14,300	0.8 ± 0.14

[a] Values present the average ± standard error (SEM) of 5–6 rabbits. [b] Mice were treated with the respective antisera 24 h prior to challenge with $2LD_{50}$ (30 µg/kg) ricin. Values present the average ± standard error (SEM) of 3 mice or 6 mice (treated with antisera harvested from rabbits immunized with native or monomerized toxin, respectively).

2.2. Characterization of the Monomerized Ricin Vaccine

2.2.1. Analytical Assessment

As the monomerized ricin antigen proved to elicit high anti-ricin antibody titers in rabbits, which conferred high level protection to ricin-intoxicated mice, we prepared deactivated monomerized ricin antigen at large amount for the immunization of a horse while carefully monitoring the various steps comprising the ricin purification and monomerization process (Figure 1A). The source material, crude ricin, displays a two-main band appearance of approximately 120 and 65 kDa on SDS-PAGE, representing *Ricinus communis* agglutinin (RCA) and ricin, respectively (lane 2), while the gel-filtration purified ricin appears as a single ~65 kDa band (lane 3) and the monomerized ricin subunits, generated by reduction and alkylation of purified ricin, appear as a ~30 kDa band doublet, representing RTA and RTB (lane 4). Next, we examined whether the purified subunits retain the RTA-dependent catalytic and RTB-dependent lectin activities. To this end, the monomerized subunit mixture was applied to an α-Lactose-Agarose column. The monomerized ricin Ultra Performance Liquid Chromatography (UPLC) chromatogram (Figure 1(B-1)), contains three main peaks, at 150, 170 and 210 s. The flow through contained only the last major peak (Figure 1(B-2)), which could be related to alkylated-RTA. The two other peaks that were attached to the column were collected only upon elution with 0.5 M galactose (Figure 1(B-3)), indicating that these two peaks represent the RTB subunit, which retains its lectin activity. To appreciate the functionality of alkylated-RTA, we assessed the activity of the alkylated-toxin in a cell free system using the reticulocyte lysate-based transcription and translation (TnT) assay. In the TnT assay (Figure 1C), wherein the toxin's activity is quantified by its ability to inhibit ribosomal translation of mRNA encoding for luciferase enzyme [14,15], not only did monomerized ricin prevent luciferase from being synthesized (ED_{50} = 0.6 ng/mL), it was also found to be a more potent inhibitor (~7 fold) than native ricin (ED_{50} = 3.9 ng/mL). Preservation of the biological activity of the monomerized ricin subunits would imply that their structural conformations did not alter in any significant manner during the monomerization process. The functional conservation of the biological activities of the two alkylated subunits may lead one to expect that antibodies raised against the toxin monomers, will interact with the holotoxin, a mandatory prerequisite for passive immunization against ricin intoxications.

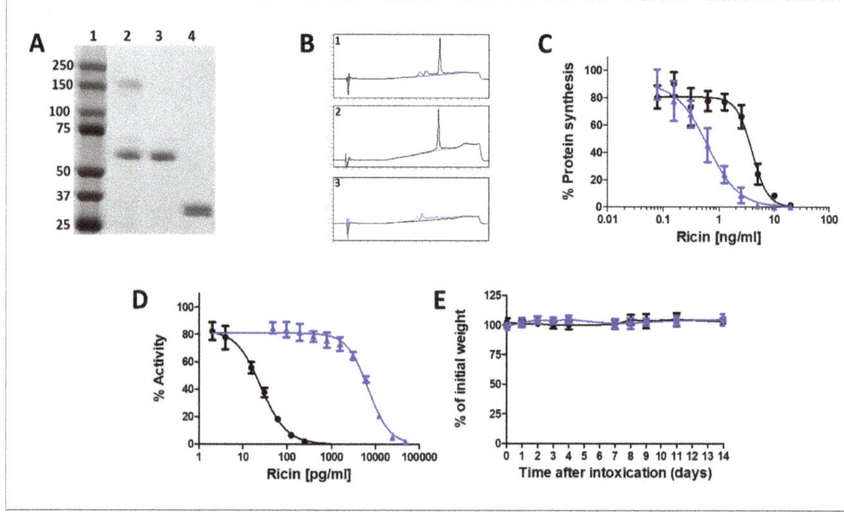

Figure 1. Characterization of the monomerized ricin vaccine. (**A**) SDS-PAGE analysis of the ricin purification and monomerization process (0.25 µg/lane) samples. Lane 1, Marker; Lane 2, Crude ricin; Lane 3, Purified ricin; Lane 4, Monomerized ricin. (**B**) UPLC chromatograms of alkylated-ricin and its purified subunits. 5 µL of the tested samples were injected into the UPLC columns (2.1 × 50 mm) and eluted at a flow rate of 0.4 mL/min. The chromatograms were monitored at 215 nm. 1. Alkylated-ricin preparation. 2. Alkylated-RTA. 3. Alkylated-RTB. (**C**) Catalytic activity assessment of ricin and monomerized ricin in a cell free system. The catalytic activities of purified ricin holotoxin (black line) and monomerized ricin (blue line) were determined using the transcription and translation (TnT) assay. Luminescence of untreated reticulocytes was considered as 100% protein synthesis. (**D**) In vitro activity of pure and alkylated ricin. Cultured HEK-293-AChE cells were incubated with increasing concentrations of pure (black line) and alkylated (blue line) ricin. The residual AChE activity in the culture medium was determined and expressed as the percent activity determined for untreated cells. (**E**) Toxicity following monomerized ricin administration to mice. Monomerized ricin (40 µg/kg body weight) was intraperitoneally administered to mice (n = 10), and body weights were determined at the indicated time points (blue line). Phosphate buffered saline (PBS) injected mice served as control (black line). Animals were observed for a 14 days period after alkylated ricin or PBS were injected.

2.2.2. In-Vitro and In-Vivo Toxicity of the Monomerized Ricin

Unlike the catalytic activity of ricin in acellular expression systems such as the TnT-based analysis described above, which is dependent on the RTA unit itself, the catalytic activity of ricin in a cell-based system, requires the presence of intact dimeric toxin molecules, since it is the lectin function of the attached RTB which enables RTA internalization into cells. To evaluate the residual cellular toxicity of the monomerized ricin, cultured HEK-293-acetylcholinesterase (AChE) cells, which produce and secrete recombinant AChE to the medium in a constitutive manner [16], were incubated with different concentrations of native or monomerized ricin, and secreted AChE levels were determined. As seen clearly (Figure 1D), protein (AChE) synthesis inhibition by monomerized ricin was dramatically reduced, the IC_{50} of monomerized ricin being 260 fold higher than that of native ricin (4729 ± 573 versus 18 ± 3 pg/mL, respectively), indicating a loss-of-activity of >99.5%. Following this in vitro assay and prior to horse immunization, we conducted an in vivo safety test in mice. To this end, 40 µg/kg monomerized-ricin (a 10-fold higher dose than the intended initial equine-vaccination dose) was injected intraperitoneally to mice, and body weights were monitored for 14 days. No body weight loss was recorded within this period of time (Figure 1E), nor were any noticeable side effects observed (data not shown).

2.3. Anti-Ricin Titer Buildup Following Horse Immunization

Although the antigen was shown to be safe, as an extra precaution, we devised a vaccination protocol for immunizing a horse, comprising a low initial dose, followed by increasing doses of the monomerized ricin antigen. Serum samples were collected three weeks after each injection, to determine ELISA and neutralizing Ab titer buildup (Figure 2). Repeated vaccinations at intervals of 3 weeks resulted in increasing antibody titers up to 11 weeks, after which antibody titers began to level off. To induce a robust immunological response, we discontinued boosting while monitoring antibody titers on a monthly basis. During a period of 3 months in which the horse was not immunized, titer levels significantly decreased, from $\sim 10^5$ to $\sim 10^4$, after which immunization was recommenced with monthly doses of 10 mg monomerized ricin. This immunization regime led to the generation of anti-ricin antibodies at titer levels that were almost one order of magnitude higher than the titer before resuming immunization. Thus, following a vaccination period of 25 weeks, high and stable ELISA- and neutralizing-antibody titers of 0.64×10^6 and 1.28×10^6 units/mL, respectively, were reached.

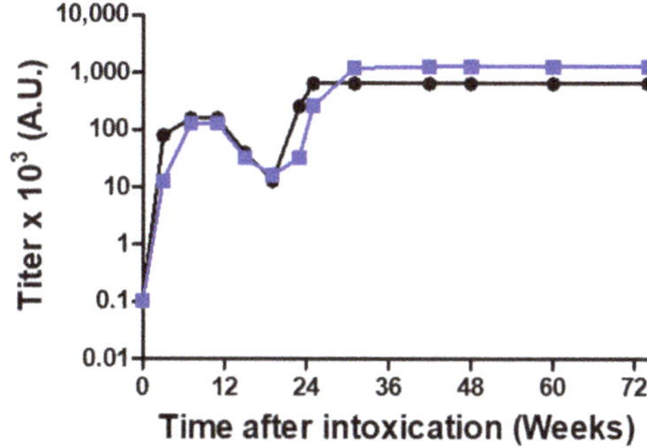

Figure 2. Titer buildup in the horse serum after vaccination with monomerized ricin immunization. The reactivity profile of the antibodies elicited by immunization were determined by enzyme linked immunosorbent assay (ELISA, black curve) and by in vitro ricin neutralization assay (blue curve).

2.4. In Vitro and In Vivo Efficacy of the Anti-Ricin Antitoxin

Pooled horse hyperimmune plasma served as the source material for the production of concentrated anti-ricin F(ab')$_2$-based antitoxin. The motivation to use a F(ab')$_2$ fragment as an antitoxin, was the potential of equine IgG antibodies to cause serum sickness [17], as the crystallizable fragment (Fc) present in IgG, may induce inflammation and unwarranted immune responses [18,19]. To characterize the F(ab')$_2$-based antitoxin, we evaluated its neutralizing potency in the cultured HEK-293-AChE cell system. To this end, ricin was mixed with increasing concentrations of anti-ricin F(ab')$_2$ and then added to the cells. As seen (Figure 3A), the antitoxin inhibited the activity of ricin and restored protein synthesis in a dose dependent manner. The antitoxin dose needed to neutralize 50% (ED$_{50}$) of the toxin was approximately ~ 0.7 nM.

To determine the efficacy of this equine-derived antitoxin, mice were intranasally challenged with a lethal dose of ricin, and 24 h later were subjected to F(ab')$_2$ antitoxin treatment via the intranasal route. Overall (Figure 3B), the treatment led to significantly high surviving ratios (62%). Since the clinical treatment of ricin-intoxicated subjects is expected to be via the intravenous route, an additional group was treated intravenously. As can be seen, the surviving ratios (65%) are practically the same as those obtained following intranasal treatment. To determine the efficacy of a late time intervention, mice were intranasally intoxicated with ricin and treated intravenously 48 h post exposure with the same horse antitoxin. Even at this late time point, considerable surviving ratios (34%) were obtained.

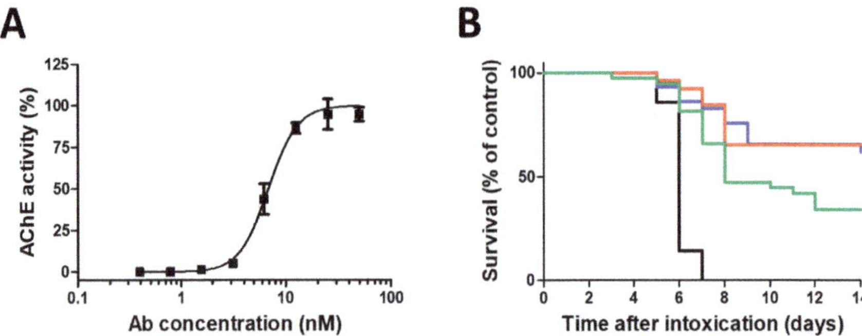

Figure 3. In vitro and in vivo efficacy of the anti-ricin F(ab')$_2$-based antitoxin. (**A**) In vitro ricin neutralization. Ricin (2 ng/mL) was mixed with increasing concentrations of the antitoxin. The mixtures were added to cultured HEK-293-AChE cells, and the residual AChE activity in the culture medium was determined 18 h later. (**B**) Kaplan–Meier survival curves of mice intoxicated with ricin and subjected to anti-ricin antibody treatment. Mice intranasally intoxicated with ricin (7 µg/kg body weight) were not treated (black line; $n = 14$), treated intranasally at 24 h post exposure (blue line; $n = 29$), treated intravenously at 24 h post exposure (red line; $n = 26$), or treated intravenously at 48 h post exposure (green line; $n = 38$) with horse derived F(ab')$_2$ anti-ricin antitoxin. Animals were observed for a 14-day period after ricin challenge.

2.5. Horse Antitoxin Attenuates Ricin-Induced Pulmonary Damage Markers

To determine the effect of horse derived F(ab')$_2$ anti-ricin antitoxin administration on pathological markers, intranasally-intoxicated mice (7 µg ricin/kg) were intravenously treated 24 h later with the antitoxin and bronchoalveolar lavage fluids (BALFs) collected at 72 h post-intoxication were analyzed for damage markers. A prominent hallmark of pulmonary ricinosis in mice is the presence of exceptionally high levels of the pro-inflammatory cytokine interleukin-6 (IL-6) in BALF [20–23]. Remarkably (Table 2), treatment with the horse derived F(ab')$_2$ anti-ricin antitoxin induced a sharp attenuation in IL-6 levels (~90% reduction); ~450 and ~3550 pg/mL of IL-6 were measured in the BALFs of antitoxin-treated and non-treated ricin-intoxicated mice, respectively.

Table 2. Pro-inflammatory and damage markers in the bronchoalveolar lavage fluids (BALFs) of mice following ricin intoxication.

	Treatment Group		
	Naïve [a]	Ricin [b]	Ricin + Antitoxin [c]
IL-6 (pg/mL)	0 ± 0	3547 ± 1372 **	451 ± 532 [&&]
Protein (mg/mL)	0.5 ± 0.1	5.7 ± 1.8 **	2.1 ± 0.8 **[&&]
ChE (mU/mL)	0 ± 0	306 ± 93 **	68 ± 31 **[&&]
XO (mU/mL)	0.6 ± 0.1	4.1 ± 1.5 **	1.7 ± 0.7 *[&&]

* $p < 0.05$ between tested group and naïve; ** $p < 0.01$ between tested group and naïve; [&&] $p < 0.01$ in comparison to ricin intoxicated mice. [a] $n = 4$; [b] $n = 6$; [c] $n = 5$. BALF samples were collected from [a] naive mice, or [b,c] ricin intoxicated mice 72 h following exposure.

Altered lung fluid balance, leading to increased permeability pulmonary edema, is a major pathophysiological characteristic of intranasal ricin intoxication and high levels of protein were reported to be present in the BALF sampled from the inflamed lungs [6,21,22], indicating that the lung–blood barrier has been disrupted. Accordingly, edema markers in mice lungs were assessed. In addition to overall protein level, we have previously demonstrated that increased pulmonary edema can be well monitored by determining cholinesterase (ChE) levels in BALF. Normally, ChE is confined to the bloodstream, yet appears at elevated levels in the BALF following disruption of the pulmonary epithelial–endothelial barrier [20]. In the antitoxin-treated mice, total protein and ChE measurements in BALF collected at 72 h post-exposure, were significantly lower in comparison to untreated mice (63% reduction in total protein, from ~5.5 to ~2 mg/mL, and 78% reduction in ChE, from ~300 to ~70 mU/mL).

Previous studies carried out in our laboratory established that xanthine oxidase (XO), an oxidative stress marker, which may also contribute to edema formation, is dramatically elevated in BALFs of mice intranasally intoxicated with ricin, at 72 h post exposure [20–22]. We therefore measured XO levels in BALFs of ricin intoxicated mice that were treated with the horse antitoxin. Indeed, XO levels were significantly reduced, by ~60% following antitoxin administration (~1.7 and ~4.0 mU/mL were measured in BALFs of antitoxin-treated and non-treated ricin-intoxicated mice, respectively) (Table 2).

3. Discussion

In the present study, a neutralized monomer-based ricin vaccine was generated by reduction and alkylation of the disulfide bond linking RTA to RTB, in order to immunize a horse for the production of highly potent neutralizing anti-ricin antibodies.

After demonstrating in rabbits that immunization with monomerized ricin was as effective as immunization with native ricin in elicitation of anti-ricin neutralizing antibodies, a thorough characterization of the antigen was conducted. Previous studies demonstrated that reduction of disulfide bonds might abolish the capability to produce biologically active antibodies, as in the case of *Plasmodium falciparum* merozoite surface glycoprotein (gp195). Reduction and alkylation of gp195 triggered an inappropriate folding of the unbound subunits, resulting in a drastic conformational change and significantly altered antigenicity [24]. Thus it was essential to demonstrate that monomerized ricin, which would be repeatedly administered to the horse over a considerably long period of time, retains its structural activity thereby implying proper folding and potential antigenicity of the irreversibly-separated subunits. Indeed, the monomeric alkylated-RTB bound firmly to an α-lactose agarose column, and eluted only following 0.5 M galactose addition, confirming RTB lectin activity. The catalytic activity of the alkylated-RTA, namely protein synthesis arrest in the TnT assay, was not only retained but even was superior to that of native RTA. This is probably due to a limited, or non-complete, separation of the holotoxin subunits of native ricin following reduction, and the relative proximity between the reduced native subunits leading to sporadic reconstitution, in contrast to the complete, irreversible, separation of the monomerized-ricin subunits.

Although the alkylated-subunits were fully active in cell free systems, a >99.5% reduction in alkylated-ricin cytotoxicity was determined in a cell culture, in which case the binding of the monomeric B subunit to the cell surface cannot promote internalization of the detached A subunit. This dramatic change in ricin-induced cytotoxicity, together with an in vivo test in mice, in which no body weight loss nor any noticeable side effect were observed, allowed the safe usage of the monomerized ricin for immunizing a horse.

Following immunization, high and stable ELISA and neutralizing antibody titers, were reached, and F(ab')$_2$-based anti-ricin antitoxin was produced from the hyperimmune horse plasma. It was previously claimed [5], that fractionation of anti-ricin IgG antibodies to F(ab')$_2$ could severely affect the neutralizing capabilities of the latter in vivo. Preliminary experiments conducted by us (data not shown) have demonstrated that the F(ab')$_2$ fragment possess the same potency as its IgG precursor both in vitro and in vivo. Indeed, the F(ab')$_2$-based antitoxin was found highly potent. The ED$_{50}$ of the antitoxin in cell culture (~0.7 nM) was comparable to the ED$_{50}$ of rabbit derived polyclonal anti-ricin IgG fraction [4]. These neutralizing capabilities at the nanomolar range are a property of highly efficient antibodies. In line with this, the antitoxin was found highly protective in mice, following administration at 24 h post intranasal intoxication with a lethal challenge of ricin. The survival ratios were >60% (whether the antitoxin was administered intranasally or intravenously), much higher than those obtained with the previously characterized rabbit-derived antibodies (~35%) [20,22]. Furthermore, high survival rates (34%) were obtained at a very late treatment time point, namely 48 h following intoxication. In sharp contrast to this, negligible surviving ratios (4%) were obtained when mice were treated with rabbit-derived anti-ricin antibody at 48 h post exposure [21]. In fact, the survival percentages of intoxicated mice treated at 24 h post exposure (PE) with horse-derived antitoxin were similar to those obtained following a combinational treatment with rabbit-derived anti-ricin antibody and immunomodulatory drugs [20,22], while the survival ratios following treatment at 48 h with horse antitoxin, were comparable to those obtained for mice treated with rabbit antibody at 24 h PE. This difference in protection is also reflected in the fact that treatment with the horse-derived F(ab')$_2$-based antitoxin induces a significant reduction in inflammatory parameters in comparison to rabbit-derived antibody-based treatment. Thus, while administration of horse antitoxin in itself induced a sharp attenuation in IL-6 levels (~90% reduction) at 72 h PE, rabbit antibody treatment had virtually no effect on IL-6 levels [25]. An additional outstanding effect of the horse antitoxin-based treatment shown in the present study was a significant reduction (by ~60%) of XO levels in the BALF of ricin-intoxicated mice. XO is an important source of reactive oxygen species (ROS) formation [26,27], which contributes significantly to pulmonary edema formation in diverse lung pathologies [28,29]. In contrast, we found that rabbit antibody-based treatment was not effective in reducing the levels of XO (data not shown). Edema markers were also dramatically reduced following treatment with the horse antitoxin (60% reduction in BALF protein content and 75% reduction in ChE levels at 72 h PE), in contrast to insignificant changes following treatment with rabbit antibody [25]. Indeed, when we previously abolished neutrophil-induced lung injury via total body irradiation (TBI), treatment at 48 h with rabbit antibody resulted in 42% survival rates, in parallel to sharp attenuation of IL-6 and edema marker levels [21].

In summary, in the present study we demonstrate that neutralized ricin antigen obtained by toxin monomerization and alkylation is completely safe for immunization, and induces high titers of potent ricin neutralizing antibodies that can be effectively used for passive immunization. F(ab')$_2$-based ricin antitoxin produced from the hyperimmune plasma of a horse that was immunized with monomerized ricin, conferred high level protection following pulmonary intoxication, which is the most fatal exposure route for ricin intoxication. This horse antitoxin-based treatment displayed effective anti-edematous and anti-inflammatory activities, apparently via balancing cytokine levels toward inflammatory attenuation in the lungs of intoxicated animals. We note that residual damage marker levels in the F(ab')$_2$-based ricin antitoxin-treated mice, nevertheless remained higher at 72 h following intoxication in comparison to the levels measured in naïve mice, suggesting that survival ratios may be further improved by combinational treatment with immunomodulatory drugs.

4. Materials and Methods

4.1. Ricin Preparation

Crude ricin was prepared from seeds of endemic *R. communis*, essentially as described before [30]. Briefly, seeds were homogenized in a Waring blender in 5% acetic acid/ PBS. The homogenate was centrifuged and the clarified supernatant containing the toxin was subjected to ammonium sulfate precipitation (60% saturation). The precipitate was dissolved in PBS and dialyzed extensively against the same buffer. The toxin preparation appeared on a Coomassie Blue stained non-reducing 10% polyacrylamide gel as 2 major bands of molecular weight approximately 65 kDa (=ricin toxin, ~80%) and 120 kDa (=ricinus communis agglutinin (RCA), ~20%). Pure toxin was prepared as described previously [30,31]. Briefly, under laminar flow and aseptic conditions the crude ricin preparation was loaded onto a gel-filtration column (Superdex 200HR 16/60 Hiload prep grade on an AKTA explorer, GE Healthcare Bio-Science AB, Uppsala, Sweden) and washed out with PBS to yield two protein peaks, corresponding to RCA and ricin. The purity of the ricin fraction was estimated by SDS-PAGE analysis to be >98%. Protein concentration was determined by 280 nm absorption (Nanodrop).

4.2. Reduction and Alkylation of Pure Ricin

Monomerized ricin vaccine was prepared, essentially as previously described [13]. Briefly, pure ricin was incubated with 50 mM Dithiothreitol (DTT, Sigma-Aldrich, Rehovot, Israel) for 2 h in room temperature, the ricin-DTT solution was incubated for additional 2 h (room temperature, protected from light) with 100 mM Iodoacetamide (IAA, Sigma-Aldrich, Rehovot, Israel), and the product (alkylated-ricin) was extensively dialyzed against PBS.

4.3. Gel Electrophoresis

Samples were visualized using Coomassie Blue stained non-reducing 10% polyacrylamide gel that was subjected to sodium dodecylsulphatepolyacrylamide gel electrophoresis (SDS-PAGE) under non-reducing conditions.

4.4. UPLC

Five μL samples (alkylated-ricin, alkylated-RTA or alkylated-RTB) were analyzed with a Water Acquity UPLC (Waters Corporation, Milford, MA, USA) equipped with a UV detector and binary solvent manager. The output signal was monitored and processed using Empower software (Empower 2.0, Waters Corporation, Milford, MA, USA). The method was employed using an Acquity UPLC BEH C-4 1.7 μm (2.1 × 50 mm) column (Waters Corporation, Milford, MA, USA). The flow rate of the mobile phase was 0.4 mL/min. The column temperature was 50 °C, and the eluted products were monitored at a wavelength of 215 nm. The samples were rinsed for 4.5 min in an acetonitrile gradient from 70% buffer A (5% acetonitrile in 0.1% trifluoroacetic acid [TFA]) and 30% buffer B (80% acetonitrile in 0.1% TFA), to 30% buffer A.

4.5. Isolation and Purification of Alkylated-Ricin Subunits

Monomerized ricin preparation was loaded on an α-Lactose-Agarose (Sigma-Aldrich, Rehovot, Israel) column, and extensively washed with PBS for collection of alkylated-RTA. Alkylated-RTB was eluted from the column with 0.5 M galactose. The purity of the isolated subunits was verified by UPLC.

4.6. Assessment of Ricin Activity in a Cell-Free Translational Assay

The biological activity of ricin was determined in a cell-free assay, as previously described [14,15]. Briefly, rabbit reticulocyte lysate containing luciferase mRNA was used to measure the activity of ricin via inhibition of protein synthesis. The relative biological activity was determined by comparing the luminescence levels in treated samples versus those of untreated controls. The amount of luciferase translated is inversely proportional to the activity of ricin.

4.7. Assessment of Ricin Activity in a Cell Culture

Genetically engineered HEK-293-acetylcholinesterase (AChE) cells [16] were cultured in Dulbecco's modified Eagle's medium (DMEM) (Biological Industries, Beit Haemek, Israel) supplemented with 10% fetal bovine serum (FBS). For the cytotoxicity studies, the cells were seeded in 96-well plates (0.75×10^5 cells/well) in medium containing different concentrations of intact or monomerized ricin. Sixteen hours later, the medium was replaced, the cells were incubated for 2 h, and the amount of secreted AChE in each well was assayed according to Ellman et al. [32] in the presence of 0.1 mg/mL bovine serum albumin (BSA), 0.3 mM 5,5′-dithiobis(2-nitrobenzoic acid), 50 mM sodium phosphate buffer (pH 8.0), and 0.5 mM acetylthiocholine iodide (ATC) (Sigma-Aldrich, Rehovot, Israel). The assay was carried out at 27 °C and monitored by a Thermomax microplate reader (Molecular Devices, Ramsey, MN, USA).

4.8. In Vitro Ricin Neutralization Assay

HEK-293-AChE cells [16] were seeded in 96-well plates (0.75×10^5 cells/well) in medium containing 2 ng/mL ricin, in the absence or presence of increasing doses of the $F(ab')_2$-based anti-ricin antitoxin. Sixteen hours later, the medium was replaced, the cells were incubated for 2 h, and the amount of secreted AChE in each well was assayed as described above.

4.9. Animal Studies

Animal experiments were performed in accordance with the Israeli law and were approved by the Ethics Committee for animal experiments at the Israel Institute for Biological Research (Horse protocol identification code: H-01-2015, date of approval: 9 August 2015; Rabbits protocol identification codes: RB-20-2011, RB-36-2012, dates of approval: 30 August 2011, 20 November 2012, respectively; Mice protocol identification code: M-59-2016, date of approval: 8 September 2016). Treatment of animals was in accordance with regulations outlined in the USDA Animal Welfare Act and the conditions specified in the Guide for Care and Use of Laboratory Animals (National Institute of Health, 1996). Local horse was immunized to produce the antitoxin preparation. New Zealand white rabbits (Charles River Laboratories Ltd., Canterbury, UK) weighing 2.5 to 3 kg were immunized in order to produce rabbit-derived polyclonal anti-ricin antibodies. Female CD-1 mice (Charles River Laboratories Ltd., UK) weighing 27–32 g were used for toxicity, survival and pathogenesis studies.

Prior to all studies in mice and rabbits, the animals were habituated to the experimental animal unit for 5 days. All mice were housed in filter-top cages in an environmentally controlled room and maintained at 21 ± 2 °C and 55 ± 10% humidity. Lighting was set to mimic a 12/12 h dawn to dusk cycle. Animals had access to food and water ad libitum.

4.10. Rabbit Anti-Ricin Hyperimmune Serum Production

Rabbits were immunized with native- or reduced- ricin in a stepwise manner, injections 1, 2, 3, and 4 containing 0.5, 4, 16, and 100 µg toxin/rabbit, with 4-week intervals between injections. Blood samples were collected (1 week after injection) to ascertain anti-ricin antibody titer build-up. Immunization was continued over 16 weeks, until steady high anti-ricin titers were observed.

4.11. Safety Studies in Mice

Mice were injected intraperitoneally with 40 µg/kg monomerized ricin, at a volume of 200 µL. Mice body weights were monitored for 14 days. In addition, careful individual detailed clinical examinations were carried out. The observations included, among others, changes in skin, fur, eyes, and occurrence of secretions and excretions. Changes in posture and autonomic activity (lacrimation and piloerection) and the presence of bizarre behavior were also checked.

4.12. In Vivo Ricin Neutralization Determination

Mice were injected intramuscularly with different volumes of rabbit-derived anti-ricin antisera, and 24 h later, the mice were intramuscularly intoxicated with a lethal dose of ricin. Preceding these studies, we determined that 15 µg crude ricin/kg body weight is approximately equivalent to 1 mouse (intramuscular) LD_{50}. Mortality was monitored over 14 days.

4.13. Survival Experiments in Mice

For intranasal intoxication, mice were anesthetized by an intraperitoneal (i.p.) injection of ketamine (1.9 mg/mouse, Vetoquinol, Lure, France) and xylazine (0.19 mg/mouse, Eurovet Animal Health, AD Bladel, The Netherlands). Crude ricin (50 µL; 7 µg/kg diluted in PBS) was applied intranasally (2×25 µL) and mortality was monitored over 14 days. Preceding these studies, we determined that 3.5 µg crude ricin/kg body weight is approximately equivalent to 1 mouse (intranasal) LD_{50}. Treatments via the intranasal route were performed on mice anesthetized as above.

For antibody treatment following intranasal exposure to ricin, anti-ricin antitoxin was delivered intranasally or intravenously at 24 or 48 h following intoxication.

4.14. Bronchoalveolar Lavage Fluid (BALF) Analysis

Mice BALF were collected by instillation of 1 mL PBS at room temperature and were centrifuged at 3000 rpm at 4 °C for 10 min. Supernatants were collected and stored at −20 °C until further use.

BALF levels of IL-6 were determined by ELISA (R&D systems, Minneapolis, MN, USA). Cholinesterase (ChE) enzymatic activity was measured according to Ellman et al. [32]. Assays were performed in the presence of 0.5 mM acetylthiocholine (Sigma-Aldrich, Rehovot, Israel), 50 mM sodium phosphate buffer pH 8.0 (Sigma-Aldrich, Rehovot, Israel), 0.1 mg/mL BSA (Sigma-Aldrich, Rehovot, Israel), and 0.3 mM 5,5'-dithiobis-(2-nitrobenzoic acid) (Sigma-Aldrich, Rehovot, Israel). The assay was carried out at 27 °C and monitored by a Thermomax microplate reader (Molecular Devices, Ramsey, MN, USA). Protein levels in BALF were determined by 280 nm absorption (NanoDrop 2000, ThermoFisher Scientific, Waltham, MA, USA). Xanthine oxidase (XO) in BALF was determined by an activity assay kit (Molecular Probes, Eugene, OR, USA).

4.15. Horse Vaccination and Plasmapheresis

A horse was immunized with escalating doses of the monomerized ricin until a minimal level of neutralizing antibodies titer is elicited. The first three doses were 2, 5 and 10 milligrams in Incomplete Freund's adjuvant (Statens Serum Institute, Copenhagen, Denmark). The following doses were adjuvant-free, at doses of 10 mg. Plasmapheresis of the hyperimmunized horse was conducted every three months using veterinary plasmapheresis instrument (plasma collection system, PCS-2, Haemonetics Corporation, Braintree, MA, USA). Ten liters of plasma were collected during each plasmapheresis procedure and plasma bags were stored at −20 °C.

4.16. F(ab')$_2$-Based Antitoxin Production

Concentrated anti-ricin F(ab')$_2$ preparations were generated from pooled horse hyperimmune antisera by one hour of pepsin (1200 U/mL, pepsin A from porcine stomach mucosa, Sigma, Steinhaim, Germany) cleavage of the Fc fragments at 30 °C and pH 3.2. The cleavage step was finalized by increasing the solution pH to 7.4 and reducing the temperature to 18 °C. Purification of the F(ab')$_2$ fragments was carried out in several stages: First, the contaminating proteins were precipitated with ammonium sulfate (25% saturation for 20 h at 18 °C). Then, the sediment that contains contaminating proteins was separated from the suspension that contains the F(ab')$_2$ fragments using crossflow microfiltration cassettes 0.2 μm (Tangenx technology, Shrewsbury, MA, USA). The filtrate of the microfiltration was washed (from small contaminating proteins and peptides), concentrated, and dialyzed (against 50 mM phosphate buffer pH 8) with crossflow ultrafiltration 30KD cassettes (Sartocon Cassette polyethersulfon 30KD, Sartorius, Goettingen, Germany). The dialyzed solution was applied on a Q-sepharose anion-exchange column (Q sepharose fast flow, GE healthcare, Uppsala, Sweden). The F(ab')$_2$ fragments fraction eluted with the flow through was collected, while the contaminating proteins remained bound to the column, which were then regenerated with 1 M NaCl. The flow through solution was adjusted to pH 6 and applied onto a SP-sepharose cation-exchange column (SP sepharose fast flow, GE healthcare, Uppsala, Sweden). The F(ab')$_2$ fragments fraction was eluted with the flow through and were collected, while, the contaminating proteins remained bound to the column, which were then regenerated with 1 M NaCl. All the chromatographic processes were carried out using an AKTA Process Instrument (GE healthcare, Uppsala, Sweden). Finally, under grade A conditions, the antitoxin solution was concentrated and dialyzed (against 300 mM glycine buffer, pH 7.4), and filtered with nanofilter (Kleenpak-Nova, PALL life science, MI, USA).

4.17. Statistical Analysis

Individual groups were compared using unpaired *t* test analysis. To estimate p values, all statistical analyses were interpreted in a two-tailed manner. Values of $p < 0.05$ were considered to be statistically significant. Kaplan–Meier analysis was performed for survival curves. All data is presented as means ± SEM.

Author Contributions: Conceptualization, R.F., A.S., T.S., C.K. and Y.G.; Methodology, R.F., A.S., R.A., M.A., S.E., A.R., A.M. (Arik Makovitzki), A.A., A.M. (Avishai Mimran), T.S., C.K. and Y.G.; Investigation, R.F., A.S., T.S., C.K. and Y.G.; Resources: R.A., M.A. and S.E.; Writing-Original Draft Preparation, R.F., A.S., T.S., C.K. and Y.G.; Visualization: R.F. and Y.G.; Project Administration, T.S., C.K. and Y.G.

Funding: This study was supported by Israel Institute for Biological Research self-funding.

Acknowledgments: We thank Eytan Elhanany for contributing to the development of the monomerized ricin vaccine preparation; Noa Caspi and Mickey Asraf for the maintenance of the horses and Moshe Mantzur and Avi Oscar for technical support.

Conflicts of Interest: The authors declare no conflict of interest.

References

1. Olsnes, S.; Kozlov, J.V. Ricin. *Toxicon* **2001**, *39*, 1723–1728. [CrossRef]
2. Greenfield, R.A.; Brown, B.R.; Hutchins, J.B.; Iandolo, J.J.; Jackson, R.; Slater, L.N.; Bronze, M.S. Microbiological, biological, and chemical weapons of warfare and terrorism. *Am. J. Med. Sci.* **2002**, *323*, 326–340. [CrossRef] [PubMed]
3. O'Hara, J.M.; Brey, R.N., 3rd; Mantis, N.J. Comparative efficacy of two leading candidate ricin toxin a subunit vaccines in mice. *Clin. Vaccine Immunol.* **2013**, *20*, 789–794. [CrossRef] [PubMed]
4. Cohen, O.; Mechaly, A.; Sabo, T.; Alcalay, R.; Aloni-Grinstein, R.; Seliger, N.; Kronman, C.; Mazor, O. Characterization and epitope mapping of the polyclonal antibody repertoire elicited by ricin holotoxin-based vaccination. *Clin. Vaccine. Immunol.* **2014**, *21*, 1534–1540. [CrossRef] [PubMed]
5. Pincus, S.H.; Das, A.; Song, K.; Maresh, G.A.; Corti, M.; Berry, J. Role of fc in antibody-mediated protection from ricin toxin. *Toxins* **2014**, *6*, 1512–1525. [CrossRef] [PubMed]
6. Pratt, T.S.; Pincus, S.H.; Hale, M.L.; Moreira, A.L.; Roy, C.J.; Tchou-Wong, K.M. Oropharyngeal aspiration of ricin as a lung challenge model for evaluation of the therapeutic index of antibodies against ricin a-chain for post-exposure treatment. *Exp. Lung. Res.* **2007**, *33*, 459–481. [CrossRef] [PubMed]
7. Vance, D.J.; Tremblay, J.M.; Mantis, N.J.; Shoemaker, C.B. Stepwise engineering of heterodimeric single domain camelid vhh antibodies that passively protect mice from ricin toxin. *J. Biol. Chem.* **2013**, *288*, 36538–36547. [CrossRef] [PubMed]
8. Maddaloni, M.; Cooke, C.; Wilkinson, R.; Stout, A.V.; Eng, L.; Pincus, S.H. Immunological characteristics associated with the protective efficacy of antibodies to ricin. *J. Immunol.* **2004**, *172*, 6221–6228. [CrossRef] [PubMed]
9. Foxwell, B.M.; Detre, S.I.; Donovan, T.A.; Thorpe, P.E. The use of anti-ricin antibodies to protect mice intoxicated with ricin. *Toxicology* **1985**, *34*, 79–88. [CrossRef]
10. Noy-Porat, T.; Rosenfeld, R.; Ariel, N.; Epstein, E.; Alcalay, R.; Zvi, A.; Kronman, C.; Ordentlich, A.; Mazor, O. Isolation of anti-ricin protective antibodies exhibiting high affinity from immunized non-human primates. *Toxins* **2016**, *8*, 64. [CrossRef] [PubMed]
11. Pincus, S.H.; Smallshaw, J.E.; Song, K.; Berry, J.; Vitetta, E.S. Passive and active vaccination strategies to prevent ricin poisoning. *Toxins* **2011**, *3*, 1163–1184. [CrossRef] [PubMed]
12. Whitfield, S.J.C.; Griffiths, G.D.; Jenner, D.C.; Gwyther, R.J.; Stahl, F.M.; Cork, L.J.; Holley, J.L.; Green, A.C.; Clark, G.C. Production, characterisation and testing of an ovine antitoxin against ricin; efficacy, potency and mechanisms of action. *Toxins* **2017**, *9*, 329. [CrossRef] [PubMed]
13. Sabo, T.; Kronman, C.; Mazor, O. Ricin-holotoxin-based vaccines: Induction of potent ricin-neutralizing antibodies. *Methods Mol. Biol.* **2016**, *1403*, 683–694. [PubMed]
14. Hale, M.L. Microtiter-based assay for evaluating the biological activity of ribosome-inactivating proteins. *Pharmacol. Toxicol.* **2001**, *88*, 255–260. [CrossRef] [PubMed]

15. Langer, M.; Rothe, M.; Eck, J.; Mockel, B.; Zinke, H. A nonradioactive assay for ribosome-inactivating proteins. *Anal. Biochem.* **1996**, *243*, 150–153. [CrossRef] [PubMed]

16. Kronman, C.; Velan, B.; Gozes, Y.; Leitner, M.; Flashner, Y.; Lazar, A.; Marcus, D.; Sery, T.; Papier, Y.; Grosfeld, H.; et al. Production and secretion of high levels of recombinant human acetylcholinesterase in cultured cell lines: Microheterogeneity of the catalytic subunit. *Gene* **1992**, *121*, 295–304. [CrossRef]

17. Bielory, L.; Kemeny, D.M.; Richards, D.; Lessof, M.H. Igg subclass antibody production in human serum sickness. *J. Allergy Clin. Immunol.* **1990**, *85*, 573–577. [CrossRef]

18. Lu, L.L.; Suscovich, T.J.; Fortune, S.M.; Alter, G. Beyond binding: Antibody effector functions in infectious diseases. *Nat. Rev. Immunol.* **2018**, *18*, 46–61. [CrossRef] [PubMed]

19. Nimmerjahn, F.; Ravetch, J.V. Fcgamma receptors as regulators of immune responses. *Nat. Rev. Immunol.* **2008**, *8*, 34–47. [CrossRef] [PubMed]

20. Gal, Y.; Mazor, O.; Alcalay, R.; Seliger, N.; Aftalion, M.; Sapoznikov, A.; Falach, R.; Kronman, C.; Sabo, T. Antibody/doxycycline combined therapy for pulmonary ricinosis: Attenuation of inflammation improved survival of ricin-intoxicated mice. *Toxicol. Rep.* **2014**, *1*, 496–504. [CrossRef] [PubMed]

21. Gal, Y.; Sapoznikov, A.; Falach, R.; Ehrlich, S.; Aftalion, M.; Kronman, C.; Sabo, T. Total body irradiation mitigates inflammation and extends the therapeutic time window for anti-ricin antibody treatment against pulmonary ricinosis in mice. *Toxins* **2017**, *9*, 278. [CrossRef] [PubMed]

22. Gal, Y.; Sapoznikov, A.; Falach, R.; Ehrlich, S.; Aftalion, M.; Sabo, T.; Kronman, C. Potent antiedematous and protective effects of ciprofloxacin in pulmonary ricinosis. *Antimicrob Agents Chemother.* **2016**, *60*, 7153–7158. [PubMed]

23. Lindauer, M.L.; Wong, J.; Iwakura, Y.; Magun, B.E. Pulmonary inflammation triggered by ricin toxin requires macrophages and il-1 signaling. *J. Immunol.* **2009**, *183*, 1419–1426. [CrossRef] [PubMed]

24. Locher, C.P.; Tam, L.Q. Reduction of disulfide bonds in plasmodium falciparum gp195 abolishes the production of growth-inhibitory antibodies. *Vaccine* **1993**, *11*, 1119–1123. [CrossRef]

25. Sabo, T.; Gal, Y.; Elhanany, E.; Sapoznikov, A.; Falach, R.; Mazor, O.; Kronman, C. Antibody treatment against pulmonary exposure to abrin confers significantly higher levels of protection than treatment against ricin intoxication. *Toxicol. Lett.* **2015**, *237*, 72–78. [CrossRef] [PubMed]

26. Komaki, Y.; Sugiura, H.; Koarai, A.; Tomaki, M.; Ogawa, H.; Akita, T.; Hattori, T.; Ichinose, M. Cytokine-mediated xanthine oxidase upregulation in chronic obstructive pulmonary disease's airways. *Pulm. Pharmacol. Ther.* **2005**, *18*, 297–302. [CrossRef] [PubMed]

27. Wright, R.M.; Ginger, L.A.; Kosila, N.; Elkins, N.D.; Essary, B.; McManaman, J.L.; Repine, J.E. Mononuclear phagocyte xanthine oxidoreductase contributes to cytokine-induced acute lung injury. *Am. J. Respir. Cell Mol. Biol.* **2004**, *30*, 479–490. [CrossRef] [PubMed]

28. Faggioni, R.; Gatti, S.; Demitri, M.T.; Delgado, R.; Echtenacher, B.; Gnocchi, P.; Heremans, H.; Ghezzi, P. Role of xanthine oxidase and reactive oxygen intermediates in lps- and tnf-induced pulmonary edema. *J. Lab. Clin. Med.* **1994**, *123*, 394–399. [PubMed]

29. Grosso, M.A.; Brown, J.M.; Viders, D.E.; Mulvin, D.W.; Banerjee, A.; Velasco, S.E.; Repine, J.E.; Harken, A.H. Xanthine oxidase-derived oxygen radicals induce pulmonary edema via direct endothelial cell injury. *J. Surg. Res.* **1989**, *46*, 355–360. [CrossRef]

30. Lin, J.Y.; Liu, S.Y. Studies on the antitumor lectins isolated from the seeds of ricinus communis (castor bean). *Toxicon* **1986**, *24*, 757–765. [CrossRef]

31. Griffiths, G.D.; Lindsay, C.D.; Upshall, D.G. Examination of the toxicity of several protein toxins of plant origin using bovine pulmonary endothelial cells. *Toxicology* **1994**, *90*, 11–27. [CrossRef]

32. Ellman, G.L.; Courtney, K.D.; Andres, V., Jr.; Feather-Stone, R.M. A new and rapid colorimetric determination of acetylcholinesterase activity. *Biochem. Pharmacol.* **1961**, *7*, 88–95. [CrossRef]

 *toxins*

Article

Characterization of MicroRNA and Gene Expression Profiles Following Ricin Intoxication

Nir Pillar, Danielle Haguel, Meitar Grad, Guy Shapira, Liron Yoffe and Noam Shomron *

Sackler Faculty of Medicine, Tel Aviv University, Tel Aviv 69978, Israel; nirpillar@gmail.com (N.P.); Haguel@mail.tau.ac.il (D.H.); meitargrad@gmail.com (M.G.); guyspersonal@gmail.com (G.S.); lironyoffe@gmail.com (L.Y.)
* Correspondence: nshomron@post.tau.ac.il; Tel.: +972-3-640-6594

Received: 25 March 2019; Accepted: 30 April 2019; Published: 2 May 2019

Abstract: Ricin, derived from the castor bean plant, is a highly potent toxin, classified as a potential bioterror agent. Current methods for early detection of ricin poisoning are limited in selectivity. MicroRNAs (miRNAs), which are naturally occurring, negative gene expression regulators, are known for their tissue specific pattern of expression and their stability in tissues and blood. While various approaches for ricin detection have been investigated, miRNAs remain underexplored. We evaluated the effect of pulmonary exposure to ricin on miRNA expression profiles in mouse lungs and peripheral blood mononuclear cells (PBMCs). Significant changes in lung tissue miRNA expression levels were detected following ricin intoxication, specifically regarding miRNAs known to be involved in innate immunity pathways. Transcriptome analysis of the same lung tissues revealed activation of several immune regulation pathways and immune cell recruitment. Our work contributes to the understanding of the role of miRNAs and gene expression in ricin intoxication.

Keywords: ricin; microRNA; lung intoxication

Key Contribution: We describe a miRNA and gene expression profile of pulmonary ricin intoxication leading to an enrichment of innate immune response pathways.

---

1. Introduction

Ricin toxin, derived from the castor bean plant Ricinus communis, is a highly toxic protein that belongs to the type 2 ribosome-inactivating protein (RIP) family [1,2]. The catalytically active subunit of ricin translocates into the cell cytoplasm where it causes irreversible inhibition of protein synthesis and ultimately cell death [3]. Several methods for developing simple, reliable, and sensitive detection approaches towards ricin have been studied [4] While many of the methods for ricin detection are robust, each has limitations in selectivity, and no single approach is currently able to fully distinguish ricin from other harmful toxins [4].

MicroRNAs (miRNAs) are small, endogenous RNA molecules, 21–24 nucleotides (nt) long. They play an important regulatory role by inducing posttranscriptional gene silencing of their mRNA targets, and thereby function as negative gene expression regulators. miRNAs contribute to the regulation of most biological processes and also influence numerous pathological states, including atherosclerosis [5], kidney diseases [6], cancer, and infectious disease [7]. Several features of miRNAs render them ideal biomarkers for toxicology. These include improved stability in biofluid samples [8], tissue specific patterns of expression [9], and highly-developed multiplexing measurement methods [10]. Although various approaches towards developing simple, reliable, and sensitive methods for ricin detection have been investigated [4,10], miRNAs have been underexplored. In this study, we aimed to examine changes in the miRNA expression profile in mice following pulmonary ricin intoxication, and the application of these changes to ricin detection. We assumed that differentially expressed miRNAs will enhance our understanding of the ricin toxicity mechanism and may be used for intoxication. We detected significant changes in miRNA expression levels following ricin exposure, including distinct enrichment of innate immunity related miRNAs. Gene expression analysis further corroborated these results, and revealed the activation of several immune regulation pathways and immune cell recruitment after ricin exposure. We believe the changes revealed following ricin exposure will set the ground for a better understanding of the intoxication process.

2. Results

To assess the effect of ricin inhalation on miRNA expression, 12 mice were intranasally challenged with a lethal dose of ricin or saline as a negative control. Twenty-four h after exposure, lungs were excised and total RNA was purified for multiplex miRNA profiling using the Nanostring nCounter system. Of the 600 mature mouse miRNAs probed, 182 had expression levels above the negative control probes (Table S1a). A total of 21 miRNAs were found to be deregulated in the ricin group compared to control samples (Table S2), 9 were significantly downregulated, and 12 were significantly upregulated, all with p-values of <0.05. Principal component analysis (PCA) using the differentially expressed (DE) miRNAs revealed sample clustering according to ricin exposure status (Figure 1). We used qPCR to validate the Nanostring results (Figure S1). To further corroborate our findings, the ricin exposure experiment and lung RNA extraction were repeated with an additional set of mice (n = 12). Five miRNAs (miR-223, miR-1224, miR-503, miR-10a, and miR-200c) exhibited statistically significant changes in expression that matched the results of the first Nanostring analysis.

Since ricin has been shown to induce rapid, massive migration of inflammatory cells—predominantly innate immune cells [11]—we utilized the innate DB database [12], which aims to capture improved coverage of the innate immunity interactome. A total of 54 immune miRNAs appear in the database. Of these, three (miR-223, miR-10a, miR-200c) were differentially expressed in our cohort. This number was significantly higher than expected by chance (p-value <0.005). To strengthen the immune regulation enrichment observation and to delineate its deregulated pathways, total RNA (the same RNA used for the first miRNA multiplex analysis) was sent for RNA-sequencing (RNA-seq). Major differences in gene expression were detected between the groups, with significant expression downregulation (p < 0.05) of 2823 genes and upregulation of 3147 genes (Table S1b). The ricin and control groups were distinctly discriminated by their gene expression (Figure S2). These vast differences in gene expression concur with previous descriptions of transcriptome analyses of models of acute lung injury mice [13]. For ontology analysis of genes that presented with statistically significant differences, high variation in expression (absolute fold-change ≥1) revealed enrichment for immune cell responses after exposure to pathogenic agents, including IL-1 and NF-kB pathway activation, leukocyte chemotaxis and migration, and chemokine activation (Figure 2). Utilizing validated miRNA–target interactions databases, we noted that many of the differentially expressed RNA-seq genes were directly regulated by the deregulated five miRNAs (Table 1).

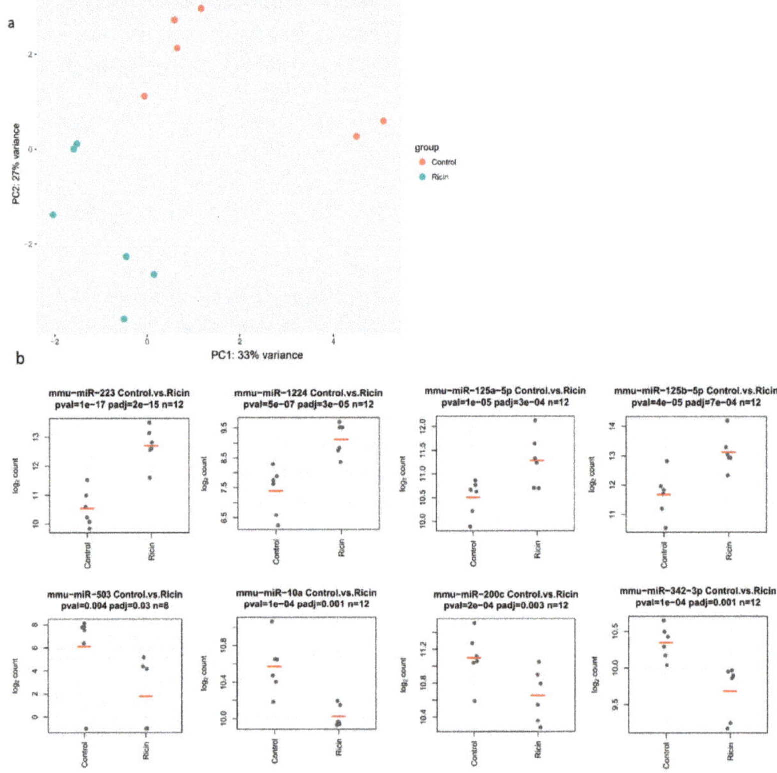

Figure 1. miRNA expression, ricin versus control treated. (**a**) PCA analysis of miRNA expression. The ricin and control groups are clearly distinguished by their miRNA expression profiles, as demonstrated by unsupervised clustering. (**b**) Expression of the top eight differentially expressed (DE) miRNAs. Each gray dot represents one sample. The red line indicates mean expression. All miRNAs had adjusted p-values < 0.05.

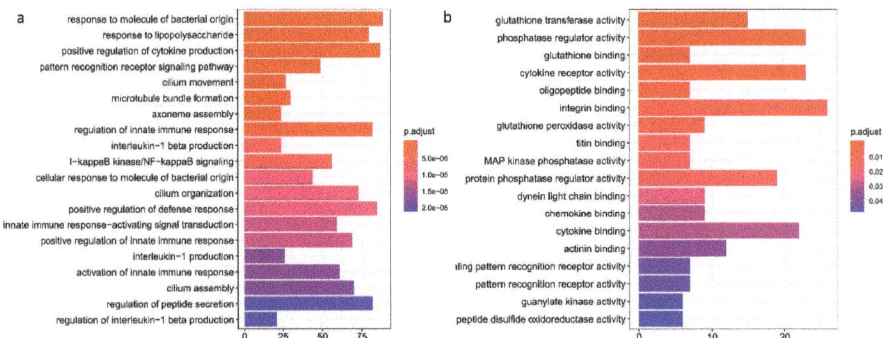

Figure 2. Gene ontology enrichment analysis of ricin intoxication. Barplot representation of the gene ontology biological function (**a**) and molecular processes (**b**) analysis of ricin induced transcriptome. The analysis comprised 2236 genes that had over 1-fold change expression between the ricin and control groups, with adjusted p-value <0.05. Both biological and molecular analyses showed enrichment for immune cell recruitment and cellular response to infection. The X-axis describes the number of genes involved in each process.

Table 1. Differentially expressed miRNAs that overlapped between both in vivo experiments (12 + 12 mice) and their differentially expressed target genes. Differentially expressed genes were taken from our RNA-sequence (seq) analysis. Validated targets were imported from miRTarBase [14].

miRNAs	Adjusted *p*-value	Mean Control Expression	Mean Ricin Expression	Differentially Expressed Target Genes
mmu-miR-223	1.75×10^{-15}	1749.90	8623.46	*Bdp1, Hpcal4, Tspyl3, Dusp8, Lif, Fmnl2, Kcnj3, Zfp467, Dbn1, Ppp4r2, F3, Pcdh17, Pvt1, Ank3, Tppp, Lonrf3, Fam13a, Ptbp2, Ankrd17, Scd1, Pdia6, Mt2, 3110043O21Rik, Rsrc2, Rest, Prkar2b, Serf1, Ntrk2, Jmjd1c, Enc1, Pitpnm3, Sgsm1, Nrep*
mmu-miR-1224	0.0000345	195.87	691.35	*Rhod, Clk4*
mmu-miR-10a	0.00136	1514.00	1318.02	*Ptpn2, Ccl9, Cenpl, Car8, Mmp25, Sgms2, Eno2, Rnd1, Tnfrsf10b, Fam227a, Tbc1d24, Gaa, Aco1, Mcc, Stam, Decr2, Rhbdl3, Ajuba, Shroom1, Ramp1, Klhl41, Ecm2*
mmu-miR-503	0.0297	190.89	23.16	*Creb5, Ncl, Inhbb*
mmu-miR-200c	0.0027	2223.93	1981.17	*Jun, Ikzf5, Map2, Sox2, Zeb2, Mgat3*

Next, we identified the diseases with over-presentation of the same highly changed genes, using disease ontology and MEDLINE/PubMed indexed articles. Enrichment in lung-related diseases of both infectious and immune related origin was detected (Figure 3). With the aim of translating these results to a clinical setting, we examined whether the ricin induced lung miRNA changes could also be detected in peripheral blood mononuclear cells (PBMCs) as a surrogate tissue that can be safely obtained from patients. PBMCs provide a large pool of gene transcripts that have demonstrated the potential to be highly sensitive to the disease microenvironments on a system-wide basis [15,16]. PBMCs were isolated from mouse blood after intranasal ricin or saline exposure. RNA was extracted and expression levels of miR-223, miR-1224, miR-10a, miR-200c, and miR-503 were evaluated using real time PCR. miR-223 was significantly upregulated in the ricin group, similar to its trend of expression in the lungs. However, other miRNAs did not present the same expression pattern (Figure 4).

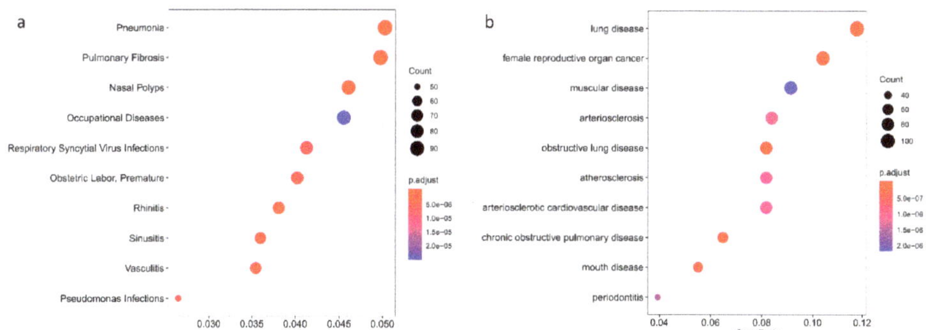

Figure 3. Disease enrichment analysis of ricin intoxication. Dotplot representation of ricin intoxication disease ontology. The analysis comprised 2,236 genes that had over 1-fold change expression between the ricin and control groups, with adjusted *p*-value < 0.05; these genes were evaluated for over-presentation in (**a**) PubMed and MEDLINE databases and (**b**) Disease Ontology. GeneRatio is the number of genes involved in the specific process divided by the total number of genes (2,236). Dot sizes ("count") represent the total number of predicted gene targets of the total number of genes that are known to be involved in the listed processes and dot color indicates statistical significance. Enrichment of lung related diseases of infectious and immune natures can be detected.

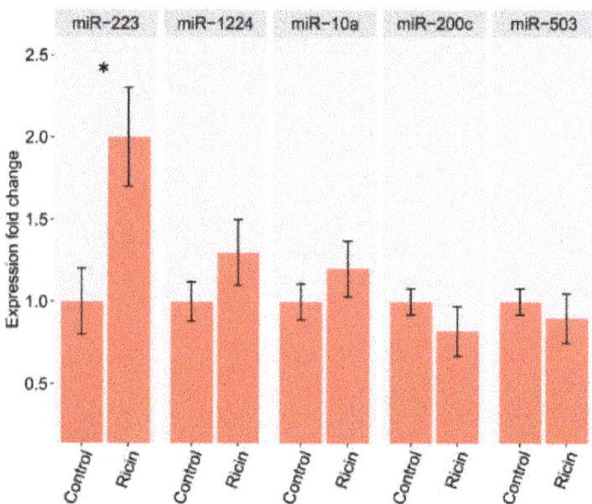

Figure 4. Relative expressions of miR-223, miR-1224, miR-10a, miR-200c, and miR-503 in peripheral blood mononuclear cells (PBMCs). Expression levels of miR-223, miR-1224, miR-10a, miR-200c, and miR-503 in PBMCs of ricin and control groups. miR-223 was found to be significantly upregulated in the ricin group, similar to its expression pattern in lung tissue. miR-1224, miR-10a, miR-200c, and miR-503 expression was not significantly different among the study groups. Data are represented as mean ± SEM. * $p < 0.05$.

Table 2. Comparison of miRNAs in studies of injured lungs.

Major miRNAs Changed	Study Description	miRNAs Overlapped with Our Study	Reference
miR-223, miR-1224, miR-503, miR-10a, miR-200c	Ricin exposure to the lungs	—	Current study
miR-142, miR-98, miR-541, miR-503, miR-653, miR- 223, miR-323, miR-196b	LPS-induced acute lung injury	miR-223, miR-503	[17]
miR-155, let-7a, let-7b, miR-125b, miR-146, miR-106a, miR-543, miR-106a, miR-7, miR-135, miR-21, miR-345, miR-223, miR-24, miR-132, miR-9, miR-503, miR-211, miR-676, let-7a, miR-200c	Ventilator-induced lung injury	miR-223, miR-503, miR-200c	[18]
miR-484, miR-425, miR-96	Mycobacterium infection (in serum)	—	[19]

3. Discussion

In the present study, we explored changes in miRNA expression profiles following pulmonary ricin intoxication. We detected 21 DE miRNAs (9 upregulated and 12 downregulated) in mouse lung tissues 24 h after ricin exposure. Five miRNAs (miR-223, miR-1224, miR-503, miR-10a, and miR-200c) had similar changes in expression in a validation cohort. Next, we utilized the InnateDB dataset, which has been developed to facilitate system level investigations of the mammalian (human, mouse, and bovine) innate immune response [12], and noted significant enrichment of the DE miRNA in the InnateDB dataset (p-value <0.005). This stands in agreement with previous studies that explored ricin-induced lung injury and innate immune response [11,20]. Associations of miR-223, miR-1224, miR-10a, miR-200c, and miR-503 with immune regulation were previously established. miR-223 was implicated in polymorphonuclear cell development and function [21]. Further, miR-223 may play a crucial role during granulopoiesis [22]; it is upregulated throughout granulocyte differentiation and is the first miRNA that was found to dramatically alter granulocyte fate [23]. miR-223 critically fine-tunes myeloid cell activity and is involved in inflammatory diseases by regulating multiple gene transcripts including E2F1, NOD-like receptor activation, and the NF-κB pathway [24]. Overexpression of miR-223 was demonstrated to dampen acute lung injury [25]. miR-1224 was previously shown to regulate tumor necrosis factor-α (TNF-α) gene activity and miR-1224 expression was demonstrated to inversely affect LPS-induced TNF-α mRNA expression [26]. Macrophages infected with mycobacterium demonstrated significantly higher miR-1224 expression, suggesting a potential role of miR-1224 in host responses upon mycobacterium infection [27]. miR-10a negatively regulates Bcl-6 expression in T cells and has an inverse effect on germinal center reactions [28]. miR-10a is involved in stabilizing the expression of Foxp3 in regulatory T cells [29] and can suppress proinflammatory monocytic cell activation by inhibiting the activation of the proinflammatory nuclear factor κB pathway [30]. miR-200c directly regulates expression of IL-6, IL-8, and CCL-5 transcripts by binding to their 3′UTRs [31]. It suppresses signaling pathways leading to NF-κB activation after TLR4 ligation; miR-200c mimics have been shown to cause decreased expression of transcripts encoding MyD88 and to induce the expression of inflammatory molecules in response to LPS [32]. Reduced miR-503 expression augments lung fibroblast VEGF production and promotes lung fibrosis [33]. Downregulation of miR-503 in bone marrow-derived mesenchymal stem cells was linked to attenuation of lung injury after infection [34], and miR-503 levels were found to be reduced in acute lung injury [35].

Lee et al. explored miRNA expression in acute lung injury (induced by LPS administration) and demonstrated significant changes in miR-223 and miR-503 expression, as well as in six other miRNAs not shared with our expression profile [17]. These discrepancies may be related to the vast differences between the ricin and LPS immune stimulation mechanisms [36]. Vaporidi et al. discovered miRNA expression profiles in ventilator-induced lung injury [18], which included changes in 65 miRNAs, among them miR-223, miR-503, and miR-200c. This abundance of deregulated miRNAs compared to our results could be explained by the pathologic differences in mechanic and pathogenic etiology of acute lung injury [37] and supports the notion that changes in miRNA expression in acute lung injury differ by their causative agents (also see Table 2).

To functionally validate the results of the miRNA expression analysis and to decipher its biological relevance, we investigated the lung transcriptome of mice exposed to ricin. Enrichment of immune regulation mediators and pathways, such as cellular response to bacterial stimulus, IL-1 and NF-kB signaling, leukocyte migration, and chemokine activity were discovered in biologic and molecular gene ontology analyses. Lung infections and immune related pathologies, such as chronic obstructive pulmonary disease, asthma, and bacterial infections were seen when utilizing the disease ontology database and PubMed indexed articles. Taken together, the miRNA expression and its transcriptome counterpart highlight the alteration in immune response following ricin intoxication.

PBMCs are a promising option for assessment of biologically distinct responses to various pathologies when the affected organs cannot be biopsied without further compromising the patient's health. Specifically, genomic analysis of PBMCs has been shown to distinguish between several lung related pathologies, including lung cancer [38], asthma [39], and pneumonia [40]. Here, we observed little resemblance between PBMC miRNA profiles and lung miRNA expression, with only the miR-223 PBMC profile mimicking its lung profile. A possible explanation for this is that most of the miRNAs expressed in ricin-induced lung toxicity originate in lung parenchyma (mostly lung epithelial cells) and the supporting stromal cells (fibroblasts, dust cells) and hence cannot be fully detected in PBMCs. Another potential explanation is that the time points used in this study are not conducive to a correlative observation between lung and PBMC miRNA responses. Future studies would benefit from a time-course assessment of miRNA profiles in response to ricin-induced intoxication.

A major limitation of our study is the usage of a rodent model to study human toxicity. Due to the complexity of running these experiments in humans, we cannot extrapolate our results to such a context. Nonetheless, we feel that our work represents an important attempt to understand the miRNAs and mRNAs modified during ricin intoxication in animals. Given that all five identified miRNAs are conserved in mammals, it is plausible that the outcome may have similarities among organisms of this class. Major strengths of our study include the comprehensive evaluation, identification, and validation of all miRNA and mRNA expression levels, and also the network and regulation analysis.

In conclusion, we describe a unique lung miRNA expression profile of pulmonary ricin intoxication that was validated in two separate in vivo experiments. The DE miRNAs detected are enriched for innate immune response and elucidate the regulatory roles of miRNAs following ricin exposure, as was further supported by transcriptome analysis. Additional studies should be conducted to further characterize the miRNA regulatory networks and to translate these findings into a miRNA-based diagnosis.

4. Materials and Methods

4.1. Animal Studies and Ricin Intoxication

Ricin preparation and animal experiments were performed at the Israel Institute for Biological Research as described previously [41]. All animal experiments were performed in accordance with Israeli regulations and were approved by the Ethics Committee for Animal Experiments at the Israel Institute for Biological Research (protocol identification code M-18-2016, date of approval: 8 September 2016). Treatment of animals was in accordance with regulations outlined in the USDA Animal Welfare Act and the conditions specified in the Guide for Care and Use of Laboratory Animals (National Institute of Health, 1996). Lung and blood samples were collected from ricin-intoxicated mice at 24 h after intranasal exposure to 7 µg/kg of crude ricin. Whole blood (0.5 mL/mouse) was extracted and treated with RBC lysis buffer (Sigma-Aldrich, St. Louis, MO, USA) followed by centrifugation of $1200 \times g$ for 20 min to isolate PBMCs. Lungs and PBMCs were stored at $-80\ ^\circ$C.

4.2. RNA Extraction

Lung tissues were homogenized using the TissueLyser LT (QIAGEN, Hilden, Germany), and total RNA from homogenized lung tissues and PBMCs were extracted using TRIzol reagent according to the manufacturer's instructions (Invitrogen, Thermo Fisher Scientific, Waltham, MA, USA). The final RNA concentration and purity were measured using a NanoDrop ND-1000 spectrophotometer (NanoDrop Technologies, Thermo Fisher Scientific).

4.3. Nanostring miRNA Expression Assay

The multiplexed NanoString nCounter miRNA expression assay (NanoString Technologies, Seattle, WA, USA) was used to profile 600 mouse miRNAs. The assay was performed according to the manufacturer's protocol. Briefly, at least 100 ng of total RNA was used as input material, with 3 µL of the threefold-diluted sample. A specific DNA tag was ligated to the 3' end of each mature miRNA, providing unique identification for each miRNA species in the sample. The tagging was performed in a multiplexed ligation reaction utilizing reverse complementary bridge oligos to achieve ligation of each miRNA to its designated tag. All hybridization reactions were incubated at 64 $^\circ$C for 18 h. Excess tags were then washed, and the resulting material was hybridized with a panel of fluorescently labeled, bar-coded reporter probes specific to the miRNA of interest. Abundances of miRNAs were quantified on the nCounter Prep Station by counting individual fluorescent barcodes and quantifying target miRNA molecules present in each sample.

4.4. Real Time PCR

The cDNA for miRNA was synthesized from total RNA. Reverse transcription of specific mature miRNAs was performed using TaqMan miRNA assays according to the manufacturer's protocol (ABI). The PCR amplification and reading were conducted in triplicate using the StepOnePlus thermal cycler (ABI). Mature miRNA expression was quantified under the following thermal cycler conditions: 2 min at 50 $^\circ$C, 10 min at 95 $^\circ$C and 40 amplification cycles (15 s at 95 $^\circ$C and 1 min at 60 $^\circ$C). Expression values were calculated based on the comparative threshold cycle (Ct) method. U6 snRNA was used for miRNA levels normalization.

4.5. RNA-Seq Preperation

RNA-seq reads were mapped to the *Mus Musculus* reference genome GRCm38 using STAR v2.4.2a [42], supplied with gene annotations downloaded from Ensembl release 82. Expression levels for each gene were quantified using HTseq-count [43]. Samples were classified as ricin or control, with 3 replicates per group. Differential expression analysis was performed using DESeq2 [44] *R* package.

4.6. Data Analysis

All statistical analyses were done using R software version 3.3. NanoString data preprocessing and normalization, followed by differential expression analysis, were performed using the DEseq2 [44] package and in-house scripts [45]. The mean value of negative controls was set as the lower threshold for each sample; only miRNAs with at least 50% of their values above the lower threshold were included in downstream analysis. miRNAs displaying absolute fold-change ≥1 with a false-discovery-rate [FDR] of 5% were considered to be differentially expressed. Gene and disease enrichment analyses were performed using the clusterProfiler [46] package. miRNA–target interaction analysis was done using the multimiR package [47].

Supplementary Materials: The following are available online at http://www.mdpi.com/2072-6651/11/5/250/s1, Figure S1: qPCR validation of Nanostring results. Figure S2: PCA analysis of mRNA expression. Table S1: (a) Expression comparison of all expressed miRNAs between ricin and control groups. (b) Expression comparison of all expressed mRNAs between ricin and control groups. Table S2: Differentially expressed miRNAs between ricin and control groups.

Author Contributions: Conceptualization, N.P. and N.S.; Methodology, N.P., L.F. and N.S.; Software, L.Y., N.P. and G.S.; Validation, D.H., M.G, L.F. and N.P.; Writing-Original Draft Preparation, N.P. and N.S.; Writing-Review & Editing, N.P. and N.S.; Visualization, N.P.; Supervision, N.S.; Funding Acquisition, N.S.

Funding: Israeli Ministry of Defense, Office of Assistant Minister of Defense for Chemical, Biological, Radiological and Nuclear (CBRN) Defense; The Edmond J. Safra Center for Bioinformatics at Tel Aviv University.

Acknowledgments: We thank investigators from the Department of Biochemistry and Molecular Genetics at the Israel Institute for Biological Studies for providing the biological specimens from intoxicated mice. We thank Avital Luba Polsky for commenting on our manuscript.

Conflicts of Interest: The authors declare no conflict of interest. The funders had no role in the design of the study; in the collection, analyses, or interpretation of data; in the writing of the manuscript, and in the decision to publish the results.

References

1. Olsnes, S.; Kozlov, J.V. Ricin. *Toxicon* **2001**, *39*, 1723–1728. [CrossRef]
2. Yermakova, A.; Mantis, N.J. Protective immunity to ricin toxin conferred by antibodies against the toxin's binding subunit (RTB). *Vaccine* **2011**, *29*, 7925–7935. [CrossRef] [PubMed]
3. Gal, Y.; Mazor, O.; Falach, R.; Sapoznikov, A.; Kronman, C.; Sabo, T. Treatments for Pulmonary Ricin Intoxication: Current Aspects and Future Prospects. *Toxins* **2017**, *9*, 331. [CrossRef] [PubMed]
4. Bozza, W.P.; Tolleson, W.H.; Rivera Rosado, L.A.; Zhang, B. Ricin detection: Tracking active toxin. *Biotechnol. Adv.* **2015**, *33*, 117–123. [CrossRef] [PubMed]
5. Novák, J.; Olejníčková, V.; Tkáčová, N.; Santulli, G. Mechanistic Role of MicroRNAs in Coupling Lipid Metabolism and Atherosclerosis. *Adv. Exp. Med. Biol.* **2015**, *887*, 79–100.
6. Metzinger-Le Meuth, V.; Metzinger, L. miR-223 and other miRNA's evaluation in chronic kidney disease: Innovative biomarkers and therapeutic tools. *Non-coding RNA Res.* **2019**, *4*, 30–35. [CrossRef]
7. Staedel, C.; Darfeuille, F. MicroRNAs and bacterial infection. *Cell Microbiol.* **2013**, *15*, 1496–1507. [CrossRef] [PubMed]
8. Cortez, M.A.; Bueso-Ramos, C.; Ferdin, J.; Lopez-Berestein, G.; Sood, A.K.; Calin, G.A. MicroRNAs in body fluids—The mix of hormones and biomarkers. *Nat. Rev. Clin. Oncol.* **2011**, *8*, 467. [CrossRef]
9. Guo, Z.; Maki, M.; Ding, R.; Yang, Y.; Zhang, B.; Xiong, L. Genome-wide survey of tissue-specific microRNA and transcription factor regulatory networks in 12 tissues. *Sci. Rep.* **2014**, *4*, 5150. [CrossRef]

10. Kolbert, C.P.; Feddersen, R.M.; Rakhsha, F.; Grill, D.E.; Simon, G.; Middha, S.; Jang, J.S.; Simon, V.; Schultz, D.A.; Zschunke, M.; et al. Multi-Platform Analysis of MicroRNA Expression Measurements in RNA from Fresh Frozen and FFPE Tissues. *PLoS ONE* **2013**, *8*, e52517. [CrossRef]

11. Lindauer, M.L.; Wong, J.; Iwakura, Y.; Magun, B.E. Pulmonary Inflammation Triggered by Ricin Toxin Requires Macrophages and IL-1 Signaling. *J. Immunol.* **2009**, *183*, 1419–1426. [CrossRef] [PubMed]

12. Breuer, K.; Foroushani, A.K.; Laird, M.R.; Chen, C.; Sribnaia, A.; Lo, R.; Winsor, G.L.; Hancock, R.E.W.; Brinkman, F.S.L.; Lynn, D.J. InnateDB: Systems biology of innate immunity and beyond—Recent updates and continuing curation. *Nucleic Acids Res.* **2013**, *41*, D1228–D1233. [CrossRef]

13. Altemeier, W.A.; Matute-Bello, G.; Gharib, S.A.; Glenny, R.W.; Martin, T.R.; Liles, W.C. Modulation of lipopolysaccharide-induced gene transcription and promotion of lung injury by mechanical ventilation. *J. Immunol.* **2005**, *175*, 3369–3376. [CrossRef] [PubMed]

14. Chou, C.-H.; Chang, N.-W.; Shrestha, S.; Hsu, S.-D.; Lin, Y.-L.; Lee, W.-H.; Yang, C.-D.; Hong, H.-C.; Wei, T.-Y.; Tu, S.-J.; et al. miRTarBase 2016: Updates to the experimentally validated miRNA-target interactions database. *Nucleic Acids Res.* **2015**, *44*, D239–D247. [CrossRef]

15. Aarøe, J.; Lindahl, T.; Dumeaux, V.; Sæbø, S.; Tobin, D.; Hagen, N.; Skaane, P.; Lönneborg, A.; Sharma, P.; Børresen-Dale, A.-L. Gene expression profiling of peripheral blood cells for early detection of breast cancer. *Breast Cancer Res.* **2010**, *12*, R7. [CrossRef]

16. Burczynski, M.E.; Dorner, A.J. Transcriptional profiling of peripheral blood cells in clinical pharmacogenomic studies. *Pharmacogenomics* **2006**, *7*, 187–202. [CrossRef]

17. Lee, W.; Kim, I.; Shin, S.; Park, K.; Yang, K.; woo Eun, J.; Sul, H.; Jeong, S. Expression profiling of microRNAs in lipopolysaccharide-induced acute lung injury after hypothermia treatment. *Mol. Cell Toxicol.* **2016**, *12*, 243–253. [CrossRef]

18. Vaporidi, K.; Vergadi, E.; Kaniaris, E.; Hatziapostolou, M.; Lagoudaki, E.; Georgopoulos, D.; Zapol, W.M.; Bloch, K.D.; Iliopoulos, D. Pulmonary microRNA profiling in a mouse model of ventilator-induced lung injury. *Am. J. Physiol. Lung Cell Mol. Physiol.* **2012**, *303*, L199–L207. [CrossRef]

19. Alipoor, S.D.; Tabarsi, P.; Varahram, M.; Movassaghi, M.; Dizaji, M.K.; Folkerts, G.; Garssen, J.; Adcock, I.M.; Mortaz, E. Serum Exosomal miRNAs Are Associated with Active Pulmonary Tuberculosis. *Dis. Markers* **2019**, *2019*, 1907426. [CrossRef] [PubMed]

20. Katalan, S.; Falach, R.; Rosner, A.; Goldvaser, M.; Brosh-Nissimov, T.; Dvir, A.; Mizrachi, A.; Goren, O.; Cohen, B.; Gal, Y.; et al. A novel swine model of ricin-induced acute respiratory distress syndrome. *Dis. Model Mech.* **2017**, *10*, 173–183. [CrossRef]

21. Johnnidis, J.B.; Harris, M.H.; Wheeler, R.T.; Stehling-Sun, S.; Lam, M.H.; Kirak, O.; Brummelkamp, T.R.; Fleming, M.D.; Camargo, F.D. Regulation of progenitor cell proliferation and granulocyte function by microRNA-223. *Nature* **2008**, *451*, 1125–1129. [CrossRef]

22. Pulikkan, J.A.; Dengler, V.; Peramangalam, P.S.; Peer Zada, A.A.; Müller-Tidow, C.; Bohlander, S.K.; Tenen, D.G.; Behre, G. Cell-cycle regulator E2F1 and microRNA-223 comprise an autoregulatory negative feedback loop in acute myeloid leukemia. *Blood* **2010**, *115*, 1768–1778. [CrossRef]

23. Mehta, A.; Baltimore, D. MicroRNAs as regulatory elements in immune system logic. *Nat. Rev. Immunol.* **2016**, *16*, 279–294. [CrossRef]

24. Neudecker, V.; Haneklaus, M.; Jensen, O.; Khailova, L.; Masterson, J.C.; Tye, H.; Biette, K.; Jedlicka, P.; Brodsky, K.S.; Gerich, M.E.; et al. Myeloid-derived miR-223 regulates intestinal inflammation via repression of the NLRP3 inflammasome. *J. Exp. Med.* **2017**, *214*, 1737–1752. [CrossRef]

25. Neudecker, V.; Brodsky, K.S.; Clambey, E.T.; Schmidt, E.P.; Packard, T.A.; Davenport, B.; Standiford, Y.S.; Weng, T.; Fletcher, A.A.; Barthel, L.; et al. Neutrophil transfer of miR-223 to lung epithelial cells dampens acute lung injury in mice. *Sci. Transl. Med.* **2017**, *9*, aah5360. [CrossRef]

26. Niu, Y.; Mo, D.; Qin, L.; Wang, C.; Li, A.; Zhao, X.; Wang, X.; Xiao, S.; Wang, Q.; Xie, Y.; et al. Lipopolysaccharide-induced miR-1224 negatively regulates tumour necrosis factor-α gene expression by modulating Sp1. *Immunology* **2011**, *133*, 8–20. [CrossRef] [PubMed]

27. Alipoor, S.D.; Mortaz, E.; Tabarsi, P.; Marjani, M.; Varahram, M.; Folkerts, G.; Garssen, J.; Adcock, I.M. miR-1224 Expression Is Increased in Human Macrophages after Infection with Bacillus Calmette-Guérin (BCG). *Iran J. Allergy Asthma Immunol.* **2018**, *17*, 250–257. [PubMed]

28. Park, H.-J.; Kim, D.-H.; Lim, S.-H.; Kim, W.-J.; Youn, J.; Choi, Y.-S.; Choi, J.-M. Insights into the role of follicular helper T cells in autoimmunity. *Immun. Netw.* **2014**, *14*, 21–29. [CrossRef] [PubMed]

29. Jeker, L.T.; Zhou, X.; Gershberg, K.; de Kouchkovsky, D.; Morar, M.M.; Stadthagen, G.; Lund, A.H.; Bluestone, J.A. MicroRNA 10a marks regulatory T cells. *PLoS ONE* **2012**, *7*, e36684. [CrossRef] [PubMed]

30. Njock, M.-S.; Cheng, H.S.; Dang, L.T.; Nazari-Jahantigh, M.; Lau, A.C.; Boudreau, E.; Roufaiel, M.; Cybulsky, M.I.; Schober, A.; Fish, J.E. Endothelial cells suppress monocyte activation through secretion of extracellular vesicles containing antiinflammatory microRNAs. *Blood* **2015**, *125*, 3202–3212. [CrossRef] [PubMed]

31. Hong, L.; Sharp, T.; Khorsand, B.; Fischer, C.; Eliason, S.; Salem, A.; Akkouch, A.; Brogden, K.; Amendt, B.A. MicroRNA-200c Represses IL-6, IL-8, and CCL-5 Expression and Enhances Osteogenic Differentiation. *PLoS ONE* **2016**, *11*, e0160915. [CrossRef] [PubMed]

32. Wendlandt, E.B.; Graff, J.W.; Gioannini, T.L.; McCaffrey, A.P.; Wilson, M.E. The role of microRNAs miR-200b and miR-200c in TLR4 signaling and NF-κB activation. *Innate Immun.* **2012**, *18*, 846–855. [CrossRef] [PubMed]

33. Ikari, J.; Nelson, A.J.; Obaid, J.; Giron-Martinez, A.; Ikari, K.; Makino, F.; Iwasawa, S.; Gunji, Y.; Farid, M.; Wang, X.; et al. Reduced microRNA-503 expression augments lung fibroblast VEGF production in chronic obstructive pulmonary disease. *PLoS ONE* **2017**, *12*, e0184039. [CrossRef]

34. Park, J.; Jeong, S.; Park, K.; Yang, K.; Shin, S. Expression profile of microRNAs following bone marrow-derived mesenchymal stem cell treatment in lipopolysaccharide-induced acute lung injury. *Exp. Ther. Med.* **2018**, *15*, 5495–5502. [CrossRef] [PubMed]

35. Ferruelo, A.; Peñuelas, Ó.; Lorente, J.A. MicroRNAs as biomarkers of acute lung injury. *Ann. Transl. Med.* **2018**, *6*, 34. [CrossRef]

36. Korcheva, V.; Wong, J.; Corless, C.; Iordanov, M.; Magun, B. Administration of ricin induces a severe inflammatory response via nonredundant stimulation of ERK, JNK, and P38 MAPK and provides a mouse model of hemolytic uremic syndrome. *Am. J. Pathol.* **2005**, *166*, 323–339. [CrossRef]

37. Beasley, M.B. The pathologist's approach to acute lung injury. *Arch. Pathol. Lab. Med.* **2010**, *134*, 719–727.

38. Showe, M.K.; Vachani, A.; Kossenkov, A.V.; Yousef, M.; Nichols, C.; Nikonova, E.V.; Chang, C.; Kucharczuk, J.; Tran, B.; Wakeam, E.; et al. Gene expression profiles in peripheral blood mononuclear cells can distinguish patients with non-small cell lung cancer from patients with nonmalignant lung disease. *Cancer Res.* **2009**, *69*, 9202–9210. [CrossRef]

39. Goleva, E.; Jackson, L.P.; Gleason, M.; Leung, D.Y.M. Usefulness of PBMCs to predict clinical response to corticosteroids in asthmatic patients. *J. Allergy Clin. Immunol.* **2012**, *129*, 687–693.e1. [CrossRef]

40. Severino, P.; Silva, E.; Baggio-Zappia, G.L.; Brunialti, M.K.C.; Nucci, L.A.; Rigato, O.; da Silva, I.D.C.G.; Machado, F.R.; Salomao, R. Patterns of gene expression in peripheral blood mononuclear cells and outcomes from patients with sepsis secondary to community acquired pneumonia. *PLoS ONE* **2014**, *9*, e91886. [CrossRef]

41. Gal, Y.; Mazor, O.; Alcalay, R.; Seliger, N.; Aftalion, M.; Sapoznikov, A.; Falach, R.; Kronman, C.; Sabo, T. Antibody/doxycycline combined therapy for pulmonary ricinosis: Attenuation of inflammation improves survival of ricin-intoxicated mice. *Toxicol. Rep.* **2014**, *1*, 496–504. [CrossRef] [PubMed]

42. Dobin, A.; Davis, C.A.; Schlesinger, F.; Drenkow, J.; Zaleski, C.; Jha, S.; Batut, P.; Chaisson, M.; Gingeras, T.R. STAR: Ultrafast universal RNA-seq aligner. *Bioinformatics* **2013**, *29*, 15–21. [CrossRef]

43. Anders, S.; Pyl, P.T.; Huber, W. HTSeq—A Python framework to work with high-throughput sequencing data. *Bioinformatics* **2015**, *31*, 166–169. [CrossRef]

44. Love, M.I.; Huber, W.; Anders, S. Moderated estimation of fold change and dispersion for RNA-seq data with DESeq2. *Genome Biol.* **2014**, *15*, 550. [CrossRef] [PubMed]

45. Pillar, N.; Bairey, O.; Goldschmidt, N.; Fellig, Y.; Rosenblat, Y.; Shehtman, I.; Haguel, D.; Raanani, P.; Shomron, N.; Siegal, T. MicroRNAs as predictors for CNS relapse of systemic diffuse large B-cell lymphoma. *Oncotarget* **2017**, *8*, 86020–86030. [CrossRef] [PubMed]

46. Yu, G.; Wang, L.-G.; Han, Y.; He, Q.-Y. ClusterProfiler: An R package for comparing biological themes among gene clusters. *OMICS* **2012**, *16*, 284–287. [CrossRef]

47. Ru, Y.; Kechris, K.J.; Tabakoff, B.; Hoffman, P.; Radcliffe, R.A.; Bowler, R.; Mahaffey, S.; Rossi, S.; Calin, G.A.; Bemis, L.; et al. The multiMiR R package and database: Integration of microRNA-target interactions along with their disease and drug associations. *Nucleic Acids Res.* **2014**, *42*, e133. [CrossRef] [PubMed]

Article

A New Method for Extraction and Analysis of Ricin Samples through MALDI-TOF-MS/MS

Roberto B. Sousa [1], **Keila S. C. Lima** [2], **Caleb G. M. Santos** [2], **Tanos C. C. França** [3,4], **Eugenie Nepovimova** [4], **Kamil Kuca** [4,*], **Marcos R. Dornelas** [2] and **Antonio L. S. Lima** [5,*]

[1]   Brazilian Army CBRN Defense Institute (IDQBRN), Avenida das Américas 28705,
     Rio de Janeiro 23020-470, Brazil; rbtsousa@gmail.com
[2]   Army Institute of Biology (IBEx), Rua Francisco Manuel, 102, Rio de Janeiro 20911-270, Brazil;
     keila@ime.eb.br (K.S.C.L.); calebguedes@gmail.com (C.G.M.S.); dornelas-ribeiro@hotmail.com (M.R.D.)
[3]   Laboratory of Molecular Modeling Applied to Chemical and Biological Defense,
     Military Institute of Engineering, Praca General Tiburcio 80, Rio de Janeiro 22290-270, Brazil;
     tanosfranca@gmail.com
[4]   Department of Chemistry, Faculty of Science, University of Hradec Kralove, Rokitanskeho 62,
     50003 Hradec Kralové, Czech Republic; eugenie.nepovimova@uhk.cz
[5]   Chemical Engineering Department, Military Institute of Engineering, Praca General Tiburcio 80,
     Rio de Janeiro 22290-270, Brazil
*   Correspondence: kamil.kuca@uhk.cz (K.K.); santoslima@ime.eb.br (A.L.S.L.);
     Tel.: +420-603-289-166 (K.K.); +55-021-2546-7050 (A.L.S.L.)

Received: 16 February 2019; Accepted: 2 April 2019; Published: 3 April 2019

Abstract: We report for the first time the efficient use of accelerated solvent extraction (ASE) for extraction of ricin to analytical purposes, followed by the combined use of sodium dodecyl sulfate-polyacrylamide gel electrophoresis (SDS-PAGE), Matrix-assisted laser desorption/ionization time-of-flight mass spectrometry (MALDI-TOF MS), and MALDI-TOF MS/MS method. That has provided a fast and unambiguous method of ricin identification for in real cases of forensic investigation of suspected samples. Additionally, MALDI-TOF MS was applied to characterize the presence and the toxic activity of ricin in irradiated samples. Samples containing ricin were subjected to ASE, irradiated with different dosages of gamma radiation, and analyzed by MALDI-TOF MS/MS for verification of the intact protein signal. For identification purposes, samples were previously subjected to SDS-PAGE, for purification and separation of the chains, followed by digestion with trypsin, and analysis by MALDI-TOF MS/MS. The results were confirmed by verification of the amino acid sequences of some selected peptides by MALDI-TOF MS/MS. The samples residual toxic activity was evaluated through incubation with a DNA substrate, to simulate the attack by ricin, followed by MALDI-TOF MS/MS analyses.

Keywords: ricin; MALDI-TOF MS; chemical weapons; biological weapons; CBRN defense

Key Contribution: ASE; SDS-PAGE; MALDI-TOF MS and MALDI-TOF MS/MS combined provide an efficient method for identification of ricin.

---

1. Introduction

Ricin is a highly toxic protein which can be extracted from the castor bean seeds (*Ricinus communis* L.). This plant is present in all Brazilian regions and explored commercially for its oil, which is mainly used for the production of lubricants, fuel and drugs. Currently, Brazil is the fourth world producer of castor bean oil, just behind India, China and Mozambique [1,2]. The production of 1.0 ton of oil generates around 1.2 ton of residue, known as castor cake [3]. The literature reports different values for the ricin content in the castor cake, varying between 0.04% and 0.08% (w/w), depending on the cultivars, the extraction method, and the analysis [4–6]. Castor cake is an excellent source of nutrients for cattle; however, its content of ricin can intoxicate the animals. In addition, the disposal of this residue in the environment represents a risk for the population. The detoxification methods proposed so far for the castor cake are expensive, time and energy demanding, and do not guarantee the total destruction of ricin without formation of other toxic products. The analyses are usually based on oral toxicity and other experiments with animals that can be influenced by several factors like species, age, and feeding time. Spectrometric techniques for identification and quantification of the products formed have rarely been used for these studies [7,8].

Ricin has toxicity similar to the neurotoxic agent sarin and can be easily extracted from the castor bean (*R. communis* L.) seeds as a fine white powder, water soluble, and stable at a large range of pH. For this reason, it is considered a chemical/biological warfare agent scheduled by both the chemical weapons convention (CWC) [9], and the biological weapons convention (BWC) [10]. It can be disseminated in the air as fine particles with a diameter smaller than 5 microns or used to contaminate water supplies or agricultural products. This turns ricin into a perfect agent for terrorist attacks and a matter of big concern for national authorities worldwide [11–14].

The structure of the ricin molecule is made up of two different chains, named RTA and RTB, connected by a disulfide bond. RTA is an N-glycosidase containing 267 amino acids arranged in eight α-helices and eight β-strands, distributed in three structural domains, forming a "U" shaped cleft containing the protein active site. RTB is a lecithin composed of 262 amino acids, containing neither α-helices nor β-strands [12,15,16]. Due to its mechanism of action in the organism, and for being a heterodimer, ricin is classified as a ribosome inactivating protein (RIP) of type II [16–18]. RTB is responsible for the binding of ricin to the terminal galactose residues of the glycolipids and glycoproteins present on the surface of eukaryotic cells [12]. This enables the formation of a vesicle surrounding the toxin, which guides it into the inner part of the cell through endocytosis. Once inside the endosome, many ricin molecules are transported back to the outside of the cell or for the lysosomes, where they are degraded. However, some of them manage to reach the Golgi complex, following in retrograde movement, until the endoplasmic reticulum, where their disulfide bonds are cleaved, splitting RTA and RTB. After, RTA is transferred to the cytosol where it reacts specifically with the ribosomal RNA (rRNA) 28S of the ribosomal subunit 60S, provoking the hydrolysis of the N-glycoside bond of the adenine residue at position 4324 (A4324) [13,19–21].

The current decontamination process of people exposed to ricin consists only on the removal of clothes, followed by washing the skin with running water [11]. In cases of ingestion, the patient should be immediately submitted to gastrointestinal lavage [22]. There is no specific antidote for poisoning with ricin yet, neither a commercial vaccine [22,23]. Many works have been performed in the last decade towards the development of a vaccine, including tests with humans. Results have been promising, however, no final product has been approved yet [22–26].

The most common methods used for detection of ricin are based on enzyme-linked immunosorbent assay, like the ELISA method, or in bioassays where the inactivation of an RNA substrate is measured. Other techniques also used are sodium dodecyl sulfate-polyacrylamide gel electrophoresis (SDS-PAGE), real-time quantitative-polymerase chain reaction (RTQ-PCR), and toxicological analyses in cell culture and in guinea pigs [7,27–33]. In this work, we report for the first time the accelerated solvent extraction (ASE) as an efficient method for ricin extraction from the seeds of castor bean (*R. communis* L.) followed by the combined use of SDS-PAGE, matrix-assisted laser desorption/ionization/time-of-flight mass spectrometry (MALDI-TOF MS) and MALDI-TOF MS/MS for a fast and unambiguous identification of ricin for forensic purposes. Additionally, this method was further successfully used to detect the presence of ricin in gamma-irradiated samples.

2. Results and Discussion

2.1. Detection of the Intact Ricin Molecule by MALDI-TOF MS

Direct analysis of the samples through MALDI-TOF MS, led to a rapid identification of the characteristic peak of the intact molecule of ricin at m/z close to 64 kD, as shown in Figure 1. The exact position of this peak can change according to the simultaneous existence of different isoforms. The non-irradiated sample (0 kGy) showed a peak relatively intense in this region of the spectra (Figure 1), while the irradiated samples showed a reduction in intensity of this peak from 5000 a.u (non-irradiated sample) to around 1500 a.u (for sample irradiated at 10 kGy), and close to 0,000 a.u (for samples irradiated at 20 and 30 kGy). Table 1 reports the ratio signal/noise (S/N) for each sample, considering the triplicates analyzed. As can be seen, the irradiation dosage of 10, 20 and 30 kGy provoked average reductions of the signal to noise (S/N) of 65.9%, 91.8% and 97.5%, respectively, compared to the non-irradiated sample.

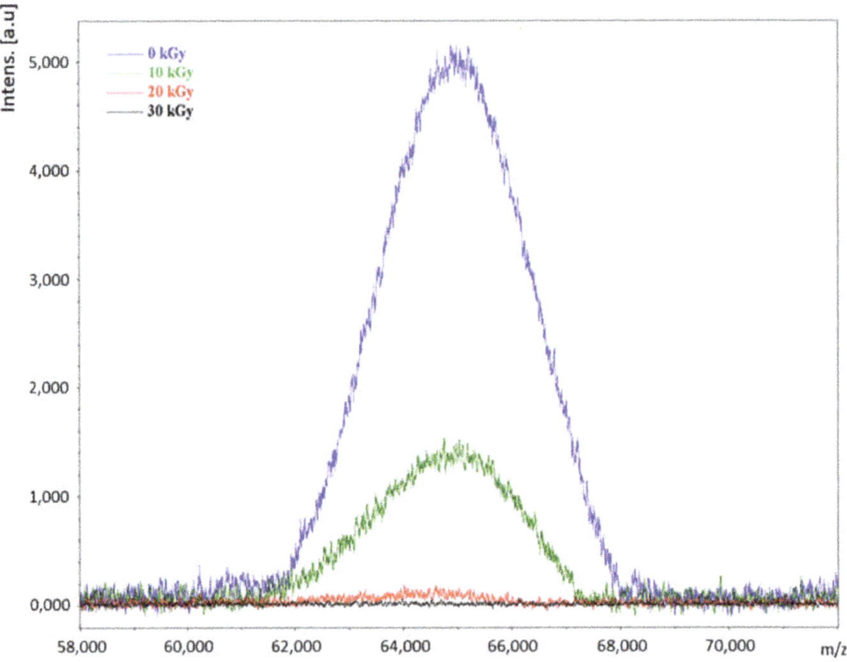

Figure 1. MALDI-TOF MS spectra of the non-irradiated ricin sample (0 kGy) and the samples irradiated at 10, 20 and 30 kGy.

Table 1. Ratio signal to noise (S/N) of ricin in the samples for each irradiation dosage (average values, three repetitions).

Irradiation Dosage (kGy)	Intensity Signal/Noise (S/N)	(S/N of Sample)/ (S/N of the Non-Irradiated Sample) (%)
0	18.1	100.00
10	6.2	34.12
20	1.5	8.19
30	0.5	2.53

Direct analysis through MALDI-TOF MS showed useful for verification of the integrity of the ricin molecules present in the samples. The presence of a peak in the range between m/z 62,000 and

68,000 should be considered as an initial detection, indicating, in a semi-quantitative way, the possible presence of non-degraded ricin. This method has the advantage of being relatively rapid.

2.2. Unequivocal Identification of Ricin

According to the procedures adopted by the Organization for Prohibition of Chemical Weapons (OPCW) [9], the unequivocal identification of a chemical substance related to the CWC should be made through two different analytical methods, and one of them must be spectrometric. In order to meet this criterion, the identification of the ricin presence in the samples was performed through MALDI-TOF MS, using of peptide mass fingerprint (PMF) method after extraction of the protein separated through SDS-PAGE. Confirmation of the results was done with MALDI-TOF MS/MS.

2.2.1. SDS-PAGE in Reducing Conditions

Samples irradiated at 0, 10, 20 and 30 kGy were submitted to SDS-PAGE in reducing conditions. This technique allowed separation of the analytes of interest from other proteins and impurities. Figure 2 shows the picture of the polyacrylamide gel from the SDS-PAGE obtained after revelation with Coomassie blue. The hydrogenation promoted by the reducing agent DTT caused the breaking of the disulfide bond between RTA and RTB. The main bands of interest were identified with numbers 1 and 2 in Figure 2 according to the crescent order of its respective molecular masses. Band 2, of higher molecular weight, presented approximately double of the length of band 1 in the vertical direction. This observation suggests the possibility of a superposition of the signals of two polypeptide chains, with similar molecular weights, at band 2.

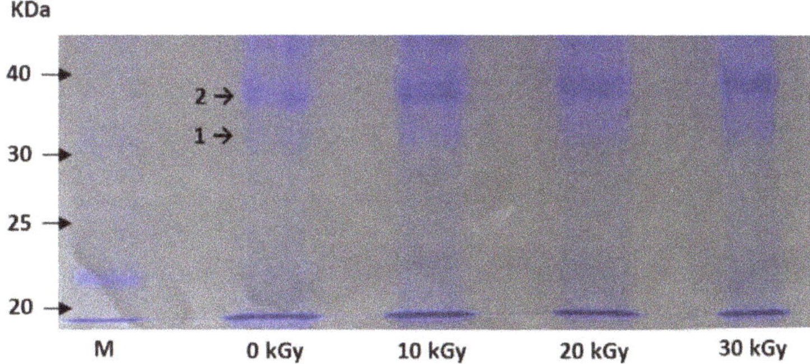

Figure 2. SDS-PAGE gel of samples irradiated at 0, 10, 20 and 30 kGy. M = marker. The bands of interest were identified with numbers 1 and 2. They were removed from the gel and identified by MALDI-TOF/MS..

One single band from SDS-PAGE may contain more than one isoform which are differentiated from each other by the degree of glycosylation because differences in the contents of sugar generate small differences in mass [34]. Fultonet et al. [35] and Kim et al. [33], observed that RTA isolated and purified exhibit two bands in the gel revealed with Coomassie blue, while RTB does not present any heterogeneity. Therefore, the whole molecule of ricin splits into two protein bands. The upper band is a mixture of RTB and the first isoform of RTA, while the lower band corresponds exclusively to the second isoform of RTA [33]. All samples presented two bands in the SDS-PAGE gel. The sample irradiated at 30 kGy presented less intense color compared to the others. Considering that the intensity is related to the amount of protein bound to the colorant used, we can deduce that this sample contains the lower concentration of ricin.

In order to make a clear identification, the two bands were collected and analyzed by MALDI-TOF MS. Results (shown in the next topic) allowed the identification of band 1 as one of the isoforms of RTA and band 2 as a mixture between RTB and the second isoform of RTA.

2.2.2. Ricin Identification through MALDI-TOF MS

The two bands in the region of interest in the SDS-PAGE gel were cut, discolored, and digested with trypsin. The peptides obtained this way were extracted from the gel and analyzed through MALDI-TOF MS. Figure 3 shows the result obtained for band 1 of the non-irradiated and the sample irradiated at 30 kGy for comparison purposes.

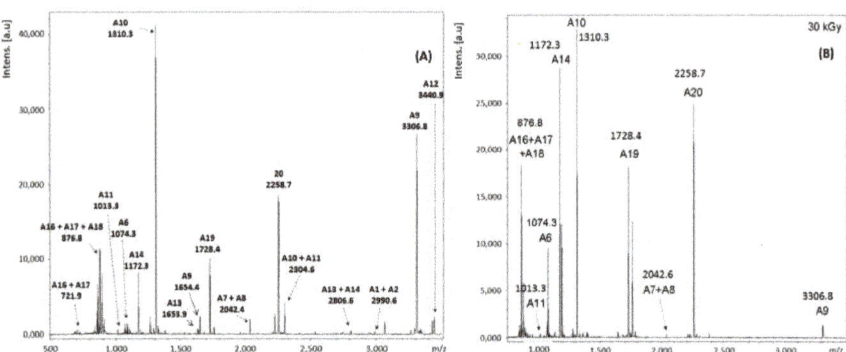

Figure 3. MALDI-TOF MS mass spectra of band 1 (RTA) of the ricin samples. (**A**) Non-irradiated and (**B**) irradiated with 30 kGy.

The mass standard spectrum, also known as the fingerprint of peptides, was used for the identification of the protein. At first, the list of the peak masses was exported to the program Biotools (Bruker®). After, through the search mechanism MASCOT PMF, the experimental results were compared to the information available in the data banks: SwissProtb [36] and National Center of Biotechnology Information (NCBI) [37]. The mass standard obtained from the analysis of band 1 showed to be similar to the chain RTA, with the probability of being a random event <0.05.

The radiolysis process may occur directly over the target molecule as a primary effect, or indirectly, through the formation and reaction of free radicals with other molecules present [38]. The gamma irradiation may denature proteins and reduce the amino acids content. Some residues, like the ones containing sulfur, can be more susceptible to radiolysis [39,40]. These phenomena may alter the intensity of peptides peaks as shown in Figure 3.

In Figure 3 the peaks were labeled in order of maintaining the correspondence with the peptides nomenclature shown in Table 2, which presents the expected masses for the complete proteolysis of ricin by trypsin. As not all theoretical scissions occurred, some peaks in Figure 3 were labeled as a sum of peptides, indicating that they remained connected. Table 2 lists the m/z values measured with the theoretical values of the corresponding peptides, and the respective amino acid sequences and the positions occupied in RTA.

Most of the peptides were identified without modifications, by the mass of their quasi-molecular ion $[M + H]^+$, like, for example, A6, A9, A10, A11, A12, A14, A19 and A20.

The presence of some peaks suggests that the proteolysis was incomplete, producing one single peptide while two or three were expected. This happened for signals at m/z A1 + A2, A7 + A8, A10 + A11, A13 + A14, A16 + A17 and A16 + A17 + A18.

As the peptide A7 + A8 has an estimated mass of 4083.214 Da, its quasi-molecular ion $[A7 + A8 + H]^+$ present m/z out of the range analyzed (700–3500) and, therefore, cannot be detected. However, it is expected that its doubled protonated ion at $m/z = 2042.615$ $[A7 + A8 + 2H]^{2+}$ be detected. This is compatible with the signal observed at $m/z = 2042.4$ (see Table 2).

Table 2. Ricin peptides identified by MALDI-TOF MS in the band 1 of the non-irradiated sample.

Peptides	Positions	Amino Acids Sequence	[M + H]$^+$ Theoretical	m/z Measured
A1 + A2	1–26	IFPKQYPIINFTTAGATVQSYTNFIR	2990.577	2990.6
A6	40–48	HEIPVLPNR	1074.605	1074.3
A7 + A8	49–85	VGLPINQRFILVELSNHAELSVTLALDVTNAYVVGYR	4084.222	2042.4 [a]
A9	86–114	AGNSAYFFHPDNQEDAEAITHLFTDVQNR	3307.504	3306.8 1654.4 [a]
A10	115–125	YTFAFGGNYDR	1310.580	1310.3
A11	126–134	LEQLAGNLR	1013.574	1013.3
A10 + A11	115–134	YTFAFGGNYDRLEQLAGNLR	2305.136	2304.6
A12	135–166	ENIELGNGPLEEAISALYYYSTGGTQLPTLAR	3440.722	3440.9
A13	167–180	SFIICIQMISEAAR	1652.849 [b]	1653.9
A13 + A14	167–189	SFIICIQMISEAARFQYIEGEMR	2806.370 [b]	2806.6
A14	181–189	FQYIEGEMR	1172.540	1172.3
A14	181–189	FQYIEGEMR	1188.535 [c]	1188.2
A16 + A17	192–196	IRYNR	721.410	721.9
A16 + A17 + A18	192–197	IRYNRR	877.512	876.8
A19	198–213	SAPDPSVITLENSWGR	1728.855	1728.4
A20	214–234	LSTAIQESNQGAFASPIQLQR	2259.173	2258.7
A23	240–258	FSVYDVSILIPIIALMVYR	2228.240 [c]	2228.5

[a] Value corresponding to the double protonated ion. [b] Considering the formation of the polyacrylamide adduct; [c] Considering oxidation of methionine to methionine sulfoxide.

The identification of peptide A13 was not trivial due to the presence of cysteine and methionine residues at positions 171 and 174, respectively. Cysteine residues can react with the acrylamide of the gel, adding 71.04 Da to the mass of the peptide. The signal observed at m/z = 1653.9 is compatible with this modification. The methionine residue may oxidize to methionine sulfoxide, originating one single peak related to A13. It was possible to identify the presence of this peak together with A14, as a low-intensity peak at 2806.6, corresponding to A13 + A14.

Peptide A14 was also associated with two other peaks at m/z 1172.2 and 1188.2 (less intense). The difference of 16 units between them can be explained due to the oxidation of the methionine residue at position 188. The same phenomenon explains the peak of peptide A23 at m/z 2228.5.

Peptides A3, A4, A15, A16, A17, A18, A21 and A22 were not detected because their masses are <700 Da, and, therefore, out of the range analyzed (700–3500).

The most intense signals of the spectra (Figure 3) were identified as A10 and A9, in descending order. This result is important because these two peptides allow differentiating ricin from the lectin RCA120 also present in castor bean samples. The similarity between the amino acids of RTA and RCA120 is superior to 93%. By comparing the RTBs of both proteins, this value drops to the still high value of 84% [41]. So, in order to eliminate doubts in the identification, it is of fundamental importance to find intense signals of some peptides that allow differentiating ricin from RCA120, like A5, A7, A9, A10, A11, A13, A22, B14, B15, B18, B19 and B20 [42].

Table 3 shows a comparison of the peptide sequences of RTA and RCA120 between positions 86 and 124. The difference between them is at amino acids 114 and 115, underlined in Table 3. While ricin holds an arginine (R) and a tyrosine (Y) at these positions, RCA120 holds a serine (S) and a phenylalanine (F). As trypsin works on the R, the cleavage happens only between amino acids 114 and 115 of ricin. Therefore, only ricin has the peptides A9 and A10, with masses 3307 and 1310 Da. The equivalent sequence of RCA120 has one single peptide with a mass of 4513 Da.

Due to the relevance of peptides A9 and A10, and the intensity of its peaks in the mass spectra of Figure 4, the ions at m/z 3307 and 1310, were chosen for confirmation of the sequence of amino acids by MALDI-TOF MS/MS as discussed in the next topic.

Table 3. Comparison between the amino acid sequences of RTA and RCA120 between positions 85 and 126 *.

Protein	Amino Acid Sequences between Positions 85 and 126 of Chain A for Ricin and RCA120
Ricin	...R[85] AGNSAYFFHPDNQEDAEAITHLFTDVQNRYTFAFGGNYDRL[126]...
RCA120	...R[85] AGNSAYFFHPDNQEDAEAITHLFTDVQNSFTFAFGGNYDRL[126]...

* Different amino acids in both sequences are underlined. The digestion with trypsin leads to peptides A9 (in red) and A10 (in blue) in ricin but keeps RCA120 as a single peptide (in green).

MALDI-TOF MS analysis of band 1 of the SDS-PAGE gel identified the presence of RTA in the non-irradiated sample. As band 1 also shows up visible in the SDS-PAGE gels of the irradiated samples, it is reasonable to suppose that RTA is also present in these samples. In fact, this was confirmed by the mass spectra shown in Figure 3 where it is possible to identify the peaks of the selected ions A9 and A10 corresponding to RTA, contrasting with a single peptide for RCA120 (see Table 3).

The main peaks found in band 1 and identified as peptides of RTA, were also observed in the spectra obtained for band 2 of the SDS-PAGE gel for the non-irradiated sample, shown in Figure 4. The main difference was the presence of peptides that compose RTB that were not observed before. Figure 4 and Table 4 present the results of the MALDI-TOF MS analysis of band 2 for the non-irradiated sample. The base ion, located at *m/z* 2230.9, corresponds to the peptide B13 (AEQQWALYADGSIRPQQNR).

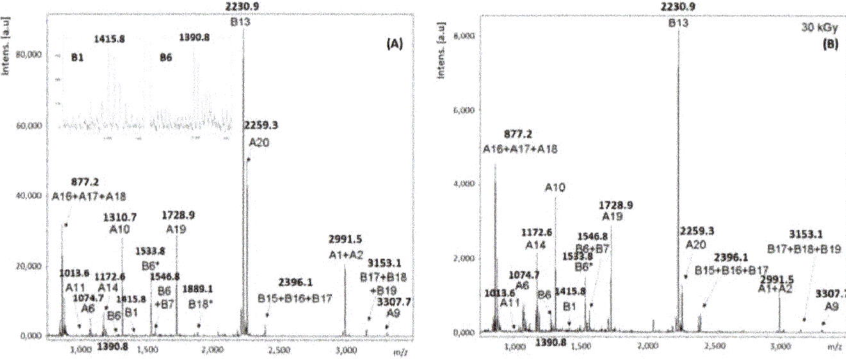

Figure 4. MALDI-TOF MS mass spectra obtained for band 2 of the SDS-PAGE gel of the ricin samples. (**A**) Non-irradiated and (**B**) irradiated with 30 kGy. Magnification of the mass spectra in the region of peptides B1 and B6 is shown in the spectra of the non-irradiated sample.

Table 4. Peptides identified by MALDI-TOF MS in band 2 of the non-irradiated sample.

Peptides	Positions	Amino Acids Sequence	[M + H]+ Theoretical	*m/z* Measured
A1 + A2	1–26	IFPKQYPIINFTTAGATVQSYTNFIR	2990.577	2991.5
A16 + A17 + A18	192–197	IRYNRR	877.512	877.2
A6	40–48	HEIPVLPNR	1074.605	1074.7
A7 + A8	49–85	VGLPINQRFILVELSNHAELSVTLALDVTNAYVVGYR	4084.222	2043.0
A9	86–114	AGNSAYFFHPDNQEDAEAITHLFTDVQNR	3307.504	3307.7
A10	115–125	YTFAFGGNYDR	1310.580	1310.7
A11	126–134	LEQLAGNLR	1013.574	1013.6
A14	181–189	FQYIEGEMR	1172.540	1172.6
A19	198–213	SAPDPSVITLENSWGR	1728.855	1728.9
A20	214–234	LSTAIQESNQGAFASPIQLQR	2259.173	2259.3
A23	240–258	FSVYDVSILIPIIALMVYR	2228.240 [a]	2229.1
B1	1–12	ADVCMDPEPIVR	1415.666	1415.8
B6	41–52	SNTDANQLWTLK	1390.696	1390.8

Table 4. *Cont.*

Peptides	Positions	Amino Acids Sequence	[M + H]$^+$ Theoretical	m/z Measured
B6 + B7	41–53	SNTDANQLWTLKR	1546.797	1546.8
B13	169–182	AEQQWALYADGSIRPQQNR	2231.095	2230.9
B15 + B16 + B17		ETVVKILSCGPASSGQRWMF K	2395.226	2396.1
B17 + B18 + B19	216–243	WMFKNDGTILNLYSGLVLDVRASDPSLK	3152.645	3153.1
B6 *	RCA120	SNTDWNQLWTLR	1533.744	1533.8
B18 *		NDGTILNLYNGLVLDVR	1889.013	1889.1

[a] Considering the oxidation of methionine to methionine sulfoxide. * Peptide of RCA120.

Besides B13, we also identified B1 and B6 among the peptides expected for RTB. These two, however, presented signals with low intensity when compared to the base ion and, therefore, are magnified in Figure 4. Other three peaks correspond to the clusters B6 + B7, B15 + B16 + B17 and B17 + B18 + B19. Peaks at m/z 1533.8 and 1889.1 do not belong to ricin but correspond to the peptides B6 * and B18 * of RCA120, being indicatives of a third component in the upper band of the SDS-PAGE gel (Figure 4 and Table 4).

The identification of the peptides present in band 2 was more difficult than for band 1. First due to the presence of different chains interfering in the spectra from each other, and increasing the complexity of the matrix. Besides, several peptides from RTB possess mass values below (B2, B4, B7, B8, B9, B15 e B17) or over (B12) the range of calibration for the method used, and, therefore, could not be identified separately. Finally, even after the addition of a reducing agent (DTT) during sample preparation, some S–S bonds do not break and others may rebind naturally. Therefore, instead of producing one single peptide, many different combinations of fragments with different masses could have happened, making it difficult the identification. A usual alternative to inhibiting the formation of new disulfide bonds after sample reduction is the addition of an alkylating agent, like iodoacetamide or iodoacetic acid, which covalently binds to the thiol group of cysteine. In this case, one should consider the increase in mass due to the addition of this new group.

Despite its major complexity related to band 1, results make it clear that band 2 is composed by the superposition of signals from RTB and one of the isoforms of RTA, corroborating with the literature [33]. Additionally, we also found evidence of the presence of peptides from RCA120. The presence of this contaminant is justified because this is a natural protein of castor bean plants with chains similar to ricin.

Results of the MALDI-TOF technique analysis of the irradiated samples of band 2 (Figure 5), were very similar to the non-irradiated sample discussed before. The same peptides were identified, and the main difference observed was the intensity reduction of the signals in the spectra compared to the non-irradiated sample. These results show that the use of the technique of MALDI-TOF after separation through SDS-PAGE allowed identification of the presence of ricin in all samples studied, including the ones irradiated at 30 kGy.

2.2.3. Analysis by MALDI-TOF MS/MS

In order to confirm the identification of ricin by a second spectrometric technique, two peptides from RTA and one from RTB were verified by MALDI-TOF MS/MS. The first and second precursor ions selected were the ones with m/z 1310 Da and 3.307 Da, due to the high intensity of its peaks in the mass spectra, and the relevance of peptides A10 and A9 for the differentiation between ricin and RCA120. The third ion was the one corresponding to B13, with m/z 2231, for being the most intense related to RTB.

The MALDI-TOF MS/MS spectra corresponding to ion at m/z 1310 is shown in Figure 5. The data obtained from this spectrum were analyzed through the software Bruker Biotools® (Version 2.2, Bruker Daltonik GmbH, Bremen, Germany), together with the search mechanism MASCOT, and compared to the data banks SwissProt [36] and NCBI [37]. Results were compatible with the amino acids sequence YTFAFGGNYDR, confirming the identification of peptide A10 from ricin.

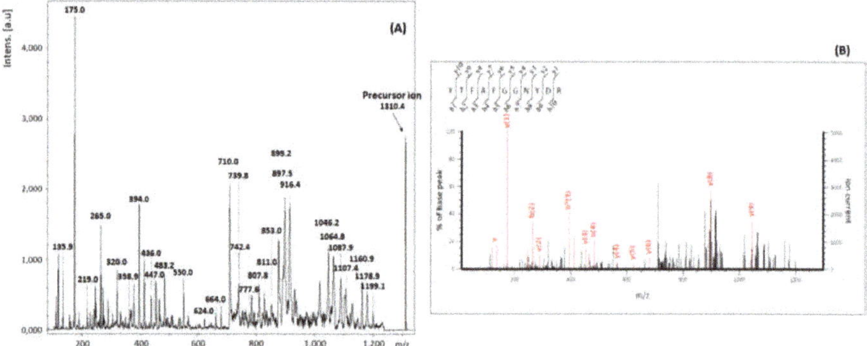

Figure 5. MALDI-TOF MS/MS spectra corresponding to the fragments of precursor ion *m/z* 1310 (**A**). Analysis of the MALDI-TOF MS/MS spectra of the precursor ion *m/z* 1310 (**B**).

Table 5 lists the fragments of peptide A10 identified by MALDI-TOF MS/MS. The following ions of series –y and –b of peptide A10, presented correspondence with the products generated from precursor *m/z* 1310: y1, y2, y3, y4, y5, y6, y7, y8 and y9; b2, b3 and b4. We also found some ions of the series a (a1, a2 and a7) and immonium, which contributed to reinforcing the interpretation of the results.

Table 5. Fragments of peptide A10 identified by MALDI-TOF MS/MS.

Ions	Amino Acid Sequences	*m/z* Theoretical	*m/z* Measured
precursor	YTFAFGGNYDR	1310.580	1310.4
b2	YT	265.118	265.0
b3	YTF	412.187	412.0
b4	YTFA	483.224	483.2
y1	R	175.119	175.0
y2	DR	290.146	290.0
y3	YDR	453.209	453.0
y4	NYDR	567.252	567.1
y5	GNYDR	624.274	623.9
y6	GGNYDR	681.295	680.8
y7	FGGNYDR	828.363	828.1
y8	AFGGNYDR	899.401	899.2
y9	FAFGGNYDR	1046.469	1046.2

Figure 6 presents the MALDI-TOF MS/MS spectra corresponding to the fragments of ion *m/z* 3307, and the corresponding analysis through the software Bruker Biotools® together with the search mechanism MASCOT, and compared to the data banks SwissProt [36] and NCBI [37]. Like before, it is possible to verify that the results are compatible with the amino acid sequence AGNSAYFFHPDNQEDAEAITHLFTDVQNR, confirming the identification of the peptide A9 of ricin. As shown in Table 6, the following series y and b were found: y1, y3, y4, y5, y6, y7, y8, y9, y11, y12, y13, y14, y15, y16, y17, y18, y20, y21, y22, y25 and y26, together with b3, b4, b7, b9, b12, b15, b23 and b25.

Finally, in order to definitely identify ricin in the samples, the last ion selected for the analysis by MALDI-TOF MS/MS was the *m/z* 2231. We tried to verify if the products formed would be compatible with the fragments of peptide B13. Results are shown in Figure 7 and Table 7.

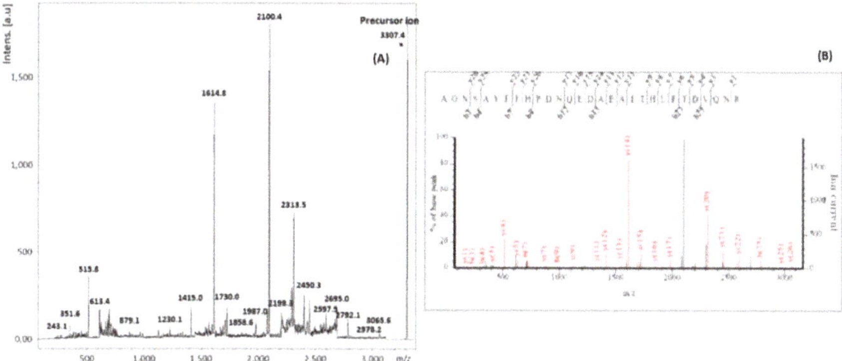

Figure 6. MALDI-TOF MS/MS spectra corresponding to the precursor ion *m/z* 3307 (**A**). Analysis of the MALDI-TOF MS/MS spectrum of the precursor ion *m/z* 3307 (**B**).

Table 6. Fragments of peptide A9 (AGNSAYFFHPDNQEDAEAITHLFTDVQNR) identified by MALDI-TOF MS/MS.

Ions	Corresponding Amino Acids Sequence	*m/z* Theoretical	*m/z* Measured
precursor	AGNSAYFFHPDNQEDAEAITHLFTDVQNR	3307.504	3307.4
b3	AGN	243.109	243.1
b4	AGNS	330.141	329.7
b7	AGNSAYF	711.310	711.3
b9	AGNSAYFFH	995.437	995.2
b12	AGNSAYFFHPDN	1321.560	1321.7
b15	AGNSAYFFHPDNQED	1693.688	1693.8
b23	AGNSAYFFHPDNQEDAEAITHLF	2576.148	2576.6
b25	AGNSAYFFHPDNQEDAEAITHLFTD	2792.222	2792.1
y1	R	175.119	175.5
y3	QNR	417.220	417.1
y4	VQNR	516.289	515.8
y5	DVQNR	631.316	631.0
y6	TDVQNR	732.363	732.0
y7	FTDVQNR	879.432	879.1
y8	LFTDVQNR	992.516	992.7
y9	HLFTDVQNR	1129.575	1129.4
y11	ITHLFTDVQNR	1343.707	1343.5
y12	AITHLFTDVQNR	1414,744	1415.0
y13	EAITHLFTDVQNR	1543.786	1544.1
y14	AEAITHLFTDVQNR	1614.823	1614.8
y15	DAEAITHLFTDVQNR	1729.850	1730.0
y16	EDAEAITHLFTDVQNR	1858.893	1858.7
y17	QEDAEAITHLFTDVQNR	1986.952	1987.0
y18	NQEDAEAITHLFTDVQNR	2100.994	2100.4
y20	PDNQEDAEAITHLFTDVQNR	2313.074	2313.5
y21	HPDNQEDAEAITHLFTDVQNR	2450.133	2450.3
y22	FHPDNQEDAEAITHLFTDVQNR	2597.202	2597.5
y25	AYFFHPDNQEDAEAITHLFTDVQNR	2978.370	2978.2
y26	SAYFFHPDNQEDAEAITHLFTDVQNR	3065.402	3065.6

The presence of ions y1, y5, y9 and y14, together with B3, B4, B6, B14 and B18, allowed confirming the similarity between the peak observed in the spectra of *m/z* 2331 and the sequence of amino acids of peptide B13 (AEQQWALYADGSIRPQQNR).

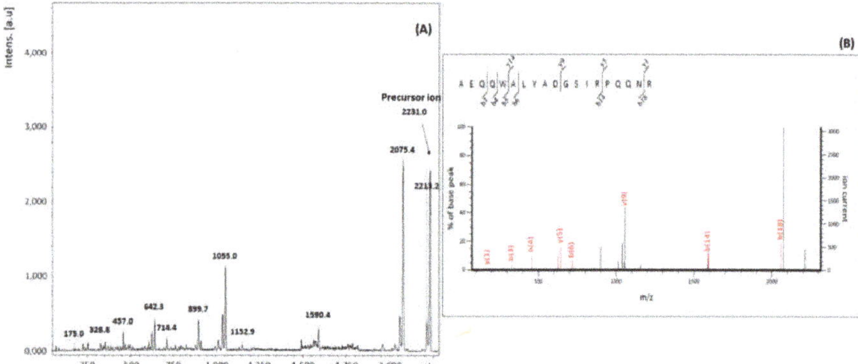

Figure 7. MALDI-TOF MS/MS spectrum corresponding to the fragments of the precursor ion *m/z* 2231. (**A**) Analysis of the MALDI-TOF MS/MS spectrum of the precursor ion *m/z* 2231 (**B**).

Table 7. Fragments of peptide B13 (AEQQWALYADGSIRPQQNR) identified by MALDI-TOF MS/MS.

Ions	Corresponding Amino Acids Sequence	*m/z* Theoretical	*m/z* Measured
precursor	AEQQWALYADGSIRPQQNR	2231.095	2231.0
b3	AEQ	329.146	328.8
b4	AEQQ	457.204	457.0
b6	AEQQWA	714.321	714.4
b14	AEQQWALYADGSIR	1589.771	1589.7
b18	AEQQWALYADGSIRPQQN	2056.984	2057.3
y1	R	175.119	175.0
y5	PQQNR	642.332	642.3
y9	GSIRPQQNR	1055.570	1055.3
y14	ALYADGSIRPQQNR	1588.819	1588.9

2.2.4. Determination of the Toxic Activity by MALDI-TOF MS

The active site responsible for the toxicity of ricin is located in RTA between residues Tyr80 and Trp211. The residues playing the most important role in the mechanism of adenine removal from rRNA 28S are Tyr80, Tyr123, Glu177 and Arg180 [20,43]. This information together with the PMF spectrum obtained by MALDI-TOF MS (Figure 3) allows correlating the ricin activity to peptides A8 to A13. Among them A8, A10 and A13 are the ones containing the most relevant residues [20,43].

All peptides in the region of the active site of ricin were identified by MALDI-TOF MS for both the non-irradiated and the irradiated samples (Figure 3). The presence of these peptides suggests the possibility of toxic activity even in the samples irradiated at 30 kGy.

In order to confirm whether the samples presented toxic activity, a non-irradiated sample and another irradiated at 30 kGy, were incubated with a buffer solution containing DNA substrate with a nucleotide sequence similar to rRNA 28S. A buffer solution containing only the DNA substrate was used as a control. Aliquots were collected at three different times of incubation (0, 4 and 24 h) and analyzed by MALDI-TOF MS. Results are shown in Figure 8. At the beginning of the reaction (letters "a", "b" and "c" in Figure 8), all samples presented a unique set of intense peaks with *m/z* values starting at 3697, followed by 3719. These spectra are compatible with the mass of the intact oligonucleotide (GCGCGAGAGCGC) (Figure 8). The first signal corresponds to the quasi-molecular ion [M + H]$^+$ and the others to adducts of salts usually present, like sodium salts [M + Na]$^+$.

After 4 h of reaction, no alteration was observed in the control sample. However, in the samples incubated with ricin (0 and 30 kGy), it was observed a peak at *m/z* 3564 with very low intensity compared to the base peak [M + H]$^+$ (Figure 8d,e,f). This same peak became much more intense in the aliquots collected after 24 h of incubation with ricin, reaching around 70% of the intensity of the base peak [M + H]$^+$ (Figure 8h,i). The control sample presented only the set [M + H]$^+$ and its adducts (Figure 8g). The difference of *m/z* between the quasi-molecular ion [M + H]$^+$ and the peak at 3564 is of 133 units. This is compatible with the replacement of one adenine base of the nucleotide sequence by a hydrogen atom. The label [M − A + H]$^+$ was used to identify this peak in Figure 8.

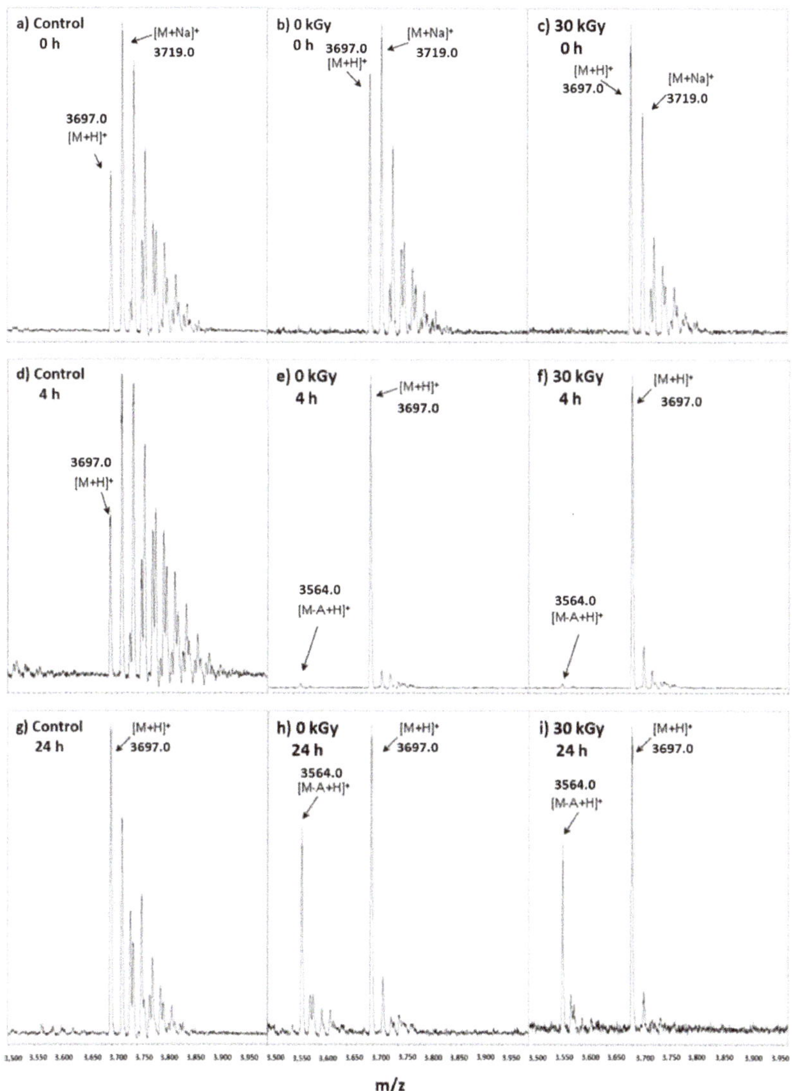

Figure 8. Verification of the toxic activity of ricin by MALDI-TOF MS. (**a**) Control at 0 h; (**b**) Sample 0 kGy at 0 h (**c**) Sample 30 kGy at 0 h; (**d**) Control after 4 h; (**e**) Sample 0 kGy after 4 h (**f**) Sample 30 kGy after 4 h; (**g**) Control after 24 h; (**h**) Sample 0 kGy after 24 h (**i**) Sample 30 kGy after 24 h.

It was possible to see by MALDI-TOF spectrometry that both samples, the non-irradiated and irradiated at 30 kGy, attacked the DNA substrate, provoking the removal of the adenine nucleotide from the sequence GCGCGAGAGCGC. This result is compatible with the MALDI-TOF MS spectra of the samples where we had already identified the presence of peptides related to the active site of ricin (Figures 3 and 4) and shows that irradiation at 30 kGy is not enough to eliminate totally the toxic activity of ricin.

3. Conclusions

Our results showed that the ASE method was efficient and rapid for the extraction of ricin samples from castor bean seeds. For the best of our knowledge, it is the first time that this method is employed to ricin extraction. This method can be improved for future works, including subsequent steps of protein purification, and comparison with other forms of sample preparation reported in the literature [4,14,32,35]. In addition, the use of ASE combined with SDS-PAGE, MALDI-TOF MS and MALDI-TOF MS/MS, has provided a fast and unambiguous identification method for ricin that can be used in real cases of forensic investigation of suspected samples.

The irradiation of samples provoked a strong and gradual reduction in the intensity of the molecular mass signal of ricin measured by MALDI-TOF MS. The signal related to the molecules that remained intact after irradiation at 30 kGy, was so small that it was not possible to distinguish it from the noise. The loss of molecular mass, however, did not imply in the complete destruction of the protein or elimination of the toxicity. Despite the initial results, ricin showed quite resistant to gamma-ray irradiation. This is illustrated by the fact that even after exposure to a dosage of 30 kGy the sample still presented toxic activity, being able to remove the adenine residue from the nucleotide sequence of the DNA substrate. These results can be attributed to two main factors: The first is related to the very low toxicity of ricin already reported by Olsnes [15] who relates that a single unit of RTA is capable of inactivating thousands of ribosomes per minute. This makes any residual remnants of ricin potentially active. The second reason is that probably the mass loss provoked by the irradiation did not alter considerably the active site of the toxin, located in a specific region of RTA. In fact, the principal trypsin peptides of the ricin chains, including the ones related to the toxic activity, were identified in all samples, including the sample irradiated at 30 kGy.

The SDS-PAGE separation technique used followed by trypsin digestion and analyzed by MALDI-TOF MS, showed an important tool of identification, in this case, making it possible to differentiate ricin from other proteins with similar structures, like the RCA120, for example. Besides, the confirmation by a second method, where the amino acids sequence of the peptides was verified by MALDI-TOF MS/MS, provided higher credibility to the identification.

4. Materials and Methods

The materials and methods used for the development of this work are described below. It is important to mention that the manipulation of ricin, even in small amounts, means a huge risk and can cause death by accidental ingestion or inhalation. Besides, the production, storage and using of this toxin are severely restricted by the CWC [9].

4.1. Protecting Equipment

The samples were produced in a glove box safety cabin equipped with a negative pressure system with HEPA and activated carbon filters, from the chemical biological radiological and nuclear (CBRN) defense Institute of the Brazilian army. Some procedures were also performed in the biology Institute of the Brazilian army in a safety cabin class II. Protective clothes, masks and gloves were needed for most of the experiments.

4.2. Sample Preparation

The scheme shown in Figure 9 summarizes all the steps used for the preparation and analysis of the ricin samples used in this work.

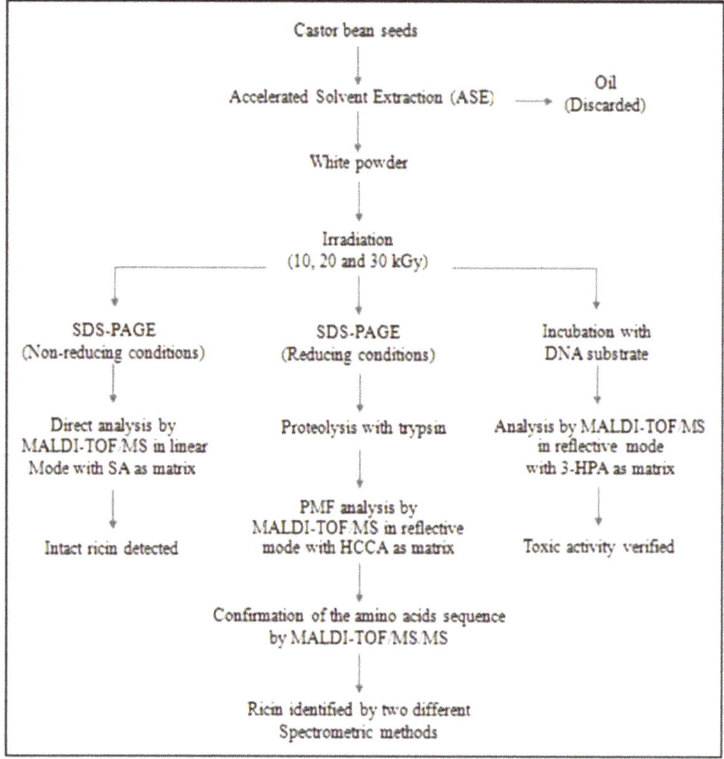

Figure 9. Scheme of preparation and analysis of the ricin samples.

4.3. Castor Bean Seeds

The castor bean seed used belonged to the species *R. communis* L., cultivar IAC Guarani, harvest 2012/2012, category S2, lot 05/2012, with 96.80% of purity. They were received from the Company "BR Seeds Production and Commerce of Seeds Ltda" based in the city of Araçatuba in the São Paulo State, Brazil.

4.4. Production of the Ricin Samples

Samples containing ricin were produced from the seeds of castor bean (*R. communis* L.) adapting the extraction method with acetone described in the literature [34,35,44], for ASE. This was performed in a Dionex extractor (Thermo Fischer Scientific, Waltham, MA, USA), model ASE100, with a 100 mL cylindrical extraction cell made of stainless steel. Firstly, the seeds were peeled with the help of tweezers and a spatula, until exposing their whitish inner part. After, they were milled in an 80 mL Ika® stainless steel knives mill, model A11 from Ika manufacturer. The oiled mass (4 g) was transferred to the extraction cell that was inserted in the oven of the extractor ASE100. The temperature was programmed to stay constant at 40 °C. Extraction was performed with a mixture 8:2 of the solvents n-hexane 99% UV/HPLC-Spectroscopic from Vetec (Rio de Janeiro, Brazil), and acetone 99.8% HPLC from J. T. Baker. This mixture was pumped into the cell, filling the whole volume, and raising the pressure to 1650 psi. After 5 min under this pressure, the extract obtained was filtered and collected into a flask. After removal of the extract, the cell was purged with nitrogen for 1 min and the extraction procedure repeated. After four rounds of extraction and evaporation of the solvent, around 1.2 g of a white powder containing ricin was separated from the extract for each extraction step.

All white powder produced was homogenized and separated in fractions of 1 g. Each fraction was packed into a 15 mL conic tube for centrifugation. After centrifugation, each tube was placed in a transparent plastic bag with a double closing system, identified externally and sent for irradiation. The samples were prepared in triplicate and named according to the irradiation dosages to be received (0 kGy, 10 kGy, 20 kGy and 30 kGy).

4.5. Samples Irradiation

The samples were irradiated with gamma rays in the installations of the CBRN Defense Institute of the Brazilian army in a research irradiator of the armored cavity type with a source of Cs^{137}, projected and constructed in 1969 in the Brookhaven National Laboratories, in the USA. The source of gamma rays consisted of 28 cylinders of CsCl, with approximately 2.5 cm of length, disposed linearly along a metallic guiding structure. The plastic bags containing the ricin samples were placed over a tray and introduced in one of the two irradiation chambers of the equipment. Samples were irradiated at dosages of 10 kGy, 20 kGy and 30 kGy.

The time of exposure to the irradiation source needed to achieve the desired dosage was calculated through software developed based on the dosimetric mapping of the irradiator. The calculations considered the current activity of the source and the density and geometry of the sample, among other factors. Equation (1) calculates the activity (A) of the source, in $kCi \cdot h^{-1}$, related to the year of the irradiation (t) [45].

$$A = 108 \cdot e^{-0.23(t - 1969)} \tag{1}$$

Based on Equation (1) the value of A for the irradiator in the year of the experiment was of $2.75 \ Ci \cdot h^{-1}$. The value found for the dosage absorbed by the samples was of $1.2 \ kGy \cdot h^{-1}$. Table 8 lists the time of irradiation necessary for achieving the dosage desired for each sample.

Table 8. Time needed for each sample.

Irradiation Dosage Absorbed	Exposure Time
10 kGy	8 h 20 min
20 kGy	16 h 40 min
30 kGy	25 h 00 min

4.6. MALDI-TOF MS Analysis

The samples were analyzed in a MALDI-TOF mass spectrometer, Bruker®, model microflex LRF. This equipment has a laser of N_2, with a maximum frequency of 60 Hz, and minimal focus of 50 µm. Each sample was analyzed in triplicate. Firstly, 0.5 µL of solution of sinapinic acid (99% from Sigma-Aldrich (São Paulo, Brazil), saturated in ethanol (95% PA from ACS, Isofar (Duque de Caxias, Rio de Janeiro, Brazil), were applied in one of the spots of a stainless steel target plate. This solution was left to dry and a thin layer of matrix was formed. After, a second solution was prepared, this time containing sinapinic acid saturated in TA30 [30% acetonitrile with 70% water/trifluoroacetic acid (99.9:0.1)]. This solution was mixed in equal parts with a third solution, containing 2 mg/mL of the original sample diluted in 0.1% TFA/water. One aliquot of 0.5 µL of this mixture was collected and applied over the first matrix layer described above, left to dry and analyzed.

The analyses were performed in linear mode by monitoring the presence of peaks in the spectral region corresponding to masses between 50 and 70 kDa. All spectra were acquired by addition after 3000 laser shots, randomly distributed over the whole surface of the sample. The laser energy was kept constant in all shots.

4.7. Inequivocal Identification of Ricin in the Samples

The identification of ricin in the samples was done through MALDI-TOF MS, using the PMF technique, after digestion of RTA and RTB.

4.7.1. SDS-PAGE Under Reducing Conditions

SDS-PAGE of gels were made in an equipment Loccus, model LP3000. The racing and stacking gels were with 12% and 5% (*w/v*) of polyacrylamide, respectively. For each sample a 20 mg/mL PBS10 buffer solution (pH 7.4) was prepared. This solution (4 μL) was added to 8 μL of a charging buffer containing the colorant bromophenol blue and other reactants (the commercial product "Blue Loading Buffer Pack", from BioLabs was used, following instructions of the manufacturer with the addition of the reducing agent dithiothreitol). Finally, 10 μL of the charging solution was applied in one of the spots of a polyacrylamide gel. The bands were removed from the SDS-PAGE gel and identified through MALDI-TOF/MS.

The electrophoresis experiments were performed with a constant tension of 200 V until the migration line achieve 1 cm from the bottom. The gel was immersed in a solution containing ethanol/glacial acetic acid/water (45:10:45) and 1 g of the colorant Coomassie blue for 30 min under gentle stirring. The revelation was done overnight with a solution of methanol/glacial acetic acid/water (3:1:6) at room temperature. Solvents and reagents used were: bright Coomassie blue R250, 98.5%, from Vetec (Rio de Janeiro, Brazil); ethanol 95%, PA, from ACS, Isofar (Duque de Caxias, Rio de Janeiro, Brazil); glacial acetic acid 99.8%, PA, from ACS, Proquímios; methanol 100%, from ACS, J.T. Baker; and distilled and deionized water produced in the lab.

4.7.2. Proteolytic Digestion

After revelation of the SDS-PAGE gel, each band related to the ricin chains was cut out with a stiletto and transferred to a 1 mL microcentrifuge tube, pre-washed twice with TA50 [50% acetonitrile with 50% water/trifluoroacetic acid (99.9:0.1)]. After, the samples were uncolored through two successive washes with 0.2 mL of a solution 100 mM of NH_4HCO_3/50% ACN, for 45 min at 37 °C. Then the samples were dehydrated by adding 100 μL of acetonitrile [99.9%, UV/HPLC spectroscopic from Vetec (Rio de Janeiro, Brazil)] for 10 min at room temperature, and dried under N_2 flow at room temperature.

In parallel, aliquots were collected from the stock solution of trypsin (trypsin gold, mass spectrometry grade from Promega (Madison, Wisconsin, USA) 1 μg/μL in 50 mM of acetic acid, and diluted to 20 μg/mL with 40 mM NH_4HCO_3/10% ACN. The dried gel pieces were then incubated and rehydrated in 30 μL of this trypsin solution at room temperature for 1 h. After, the digestion buffer (40 mM NH_4HCO_3/10% ACN) was added until covering completely the gel pieces. The tubes were well closed to avoid evaporation and incubated overnight at 37 °C. The day after the solution was transferred to a clean tube and 30 μL of TA50 added to the gel, which was submitted to ultrasound for 20 minutes. The resulting solution was transferred to a clean tube and totally dried under N_2 (AP, 99.997% from Linde) flow, being re-suspended again with 20 μL of the solution of TFA 0.1% in water. This sample, containing the peptides from the trypsin digestion was sent for analysis by MALDI-TOF MS.

Table 9 shows the peptides expected for the complete trypsinization of ricin, named according to its positions in the sequences of RTA and RTB.

Table 9. Expected peptides from the total proteolysis of ricin with trypsin.

Abbreviation	Position	Sequence of Amino Acids	Molecular Mass (M)
A1	36–39	IFPK	504.3
A2	40–61	QYPIINFTTAGATVQSYTNFIR	2504.3
A3	62–64	AVR	344.2

Table 9. *Cont.*

Abbreviation	Position	Sequence of Amino Acids	Molecular Mass (M)
A4	65–66	GR	231.1
A5	67–74	LTTGADVR	831.4
A6	75–83	HEIPVLPNR	1073.6
A7	84–91	VGLPINQR	895.5
A8	92–120	FILVELSNHAELSVTLALDVTNAYVVGYR	3205.7
A9	121–149	AGNSAYFFHPDNQEDAEAITHLFTDVQNR	3306.5
A10	150–160	YTFAFGGNYDR	1309.6
A11	161–169	LEQLAGNLR	1012.6
A12	170–201	ENIELGNGPLEEAISALYYYSTGGTQLPTLAR	3439.7
A13	202–215	SFIICIQMISEAAR	1580.8
A14	216–224	FQYIEGEMR	1171.5
A15	225–226	TR	275.2
A16	227–228	IR	287.2
A17	229–231	YNR	451.2
A18	232–232	R	174.1
A19	233–248	SAPDPSVITLENSWGR	1727.9
A20	249–269	LSTAIQESNQGAFASPIQLQR	2258.2
A21	270–270	R	174.1
A22	271–274	NGSK	404.2
A23	275–293	FSVYDVSILIPIIALMVYR	2211.2
A24	294–302	CAPPPSSQF	932.4
B1	315–326	ADVCMDPEPIVR	1343.6
B2	327–330	IVGR	443.3
B3	331–338	NGLCVDVR	874.4
B4	339–341	DGR	346.2
B5	342–354	FHNGNAIQLWPCK	1526.8
B6	355–366	SNTDANQLWTLK	1389.7
B7	367–367	R	174.1
B8	368–372	DNTIR	617.3
B9	373–376	SNGK	404.2
B10	373–376	CLTTYGYSPGVYVMIYDCNTAATDATR	2948.3
B11	404–416	WQIWDNGTIINPR	1611.8
B12	417–482	SSLVLAATSGNSGTTLTVQTNIYAVSQGWLPT NNTQPFVTTIVGLYGLCLQANSGQVWIED CSSEK	6932.4
B13	417–482	AEQQWALYADGSIRPQQNR	2230.1
B14	502–512	DNCLTSDSNIR	1236.5
B15	513–517	ETVVK	574.3
B16	518–529	ILSCGPASSGQR	1174.6
B17	530–533	WMFK	610.3
B18	534–550	NDGTILNLYSGLVLDVR	1861.0
B19	551–557	ASDPSLK	716.4
B20	558–576	QIILYPLHGDPNQIWLPLF	2276.2

4.7.3. MALDI-TOF MS Analysis for the Identification of Ricin

The mixture of trypsin peptides extracted from each band of the SDS-PAGE gel was analyzed through MALDI-TOF MS using a saturated solution of 4-hydroxy-α-cyanocinnamic acid (HCCA) saturated in TA30 as a matrix. The sample preparation consisted of mixing equal volumes of the peptides solution in 0.1% TFA/water with a saturated solution of HCCA in TA30. After, 0.5 μL of this new mixture was applied over the target plate and left to dry. The analyses were performed in reflector mode, by monitoring the presence of peaks in the spectral region corresponding to the weight between 700 and 4000 Da. All spectra were obtained by addition of 2000 laser shots randomly distributed over the whole surface of the sample. The laser energy was kept constant during the shots.

The mass spectra of the mixture of peptides were used for the identification of ricin. Firstly, the *m/z* list of the peaks obtained was exported for the software Biotools, from Bruker. After, through the MASCOT PMF search mechanism, the experimental results were compared with the information available in the data banks of proteins SwissProt [36] and NCBI [37].

4.7.4. Analyses by MALDI-TOF MS/MS

The results obtained by MALDI-TOF MS were confirmed by a second analytic spectrometric technique. For this, three peptides were chosen to have their amino acid sequences verified through MALDI-TOF MS/MS. The criteria established for the selection of the precursor ions were the intensity and the relevance of the peptide for the differentiation between ricin and other proteins, the absence of cysteine and methionine residues, and the possibility of the same peptide representing RTA and RTB.

The analyses were performed in the same target plate with the samples former prepared for the MALDI-TOF MS experiment. The equipment used also was the same used before. The spectra were obtained by the method known as fragmentation analysis and structural time of flight (FAST), which only works in the reflective mode. For each analysis the range of the ions selector and the number of segments were adjusted according to the mass of the precursor ion selected, avoiding interference of fragments from possible adjacent ions. The spectra were exported to the software Bruker Biotools and, with the help of the MASCOT searching mechanism, compared to the data existing in the data banks SwissProt [36] and NCBI [37].

4.8. Verification of the Toxic Activity of the Ricin Samples by MALDI-TOF MS

The toxic activity of ricin present in the samples was verified by MALDI-TOF MS, following a method adapted from Schieltz et al. [46]. For this, a DNA substrate chemically synthesized with the nucleotide sequence GCGCGAGAGCGC, similar to rRNA 28S where the ricin attack occurs, was acquired from the Company Genone Biotechnologies (Rio de Janeiro, Brazil).

A solution containing 0.1 µmol/mL of nucleotides was prepared and mixed with a PBS10 buffer (pH 7.4). After, the sample solution was prepared with 20 mg/mL of the white powder extracted from castor bean seeds mixed with the PBS10 buffer solution. The reaction mixture was produced by mixing equal volumes of the two solutions. Then it was incubated at 37 °C, without stirring, for 24 h. Aliquots were collected and analyzed in times 0, 4 and 24 h. The matrix solution consisted of 3-hydroxypicolinic acid (3-HPA) saturated in TA50. This solution (0.5 µL) was applied over the target plate and left to dry at room temperature. At the time intervals mentioned above, 2.0 µL were collected from the supernatant of the reaction and mixed with more 18 µL of the matrix solution. From this mixture, one aliquot of 0.5 µL was deposited over the first layer, left for drying, and introduced in the target plate of the equipment.

The analysis method was in the reflective mode, with a range of *m/z* from 3000 to 4000, with the addition of spectra obtained after 2000 laser shots randomly distributed over the whole sample surface in the target plate.

We monitored the intensities of the signals of peaks at *m/z* 3,697, referring to the mass of the quasi-molecular ion of the oligonucleotide protonated $[M + H]^+$ and in *m/z* 3564, related to the loss of adenine $[M+H-A]^+$.

Solvents and reagents used for these experiments were: 3-hydroxypicolinic acid 99%, from Sigma Aldrich, TA50 (produced with acetonitrile 99,9%, UV/HPLC spectroscopic, from Vetec (Rio de Janeiro, Brazil); trifluoroacetic acid 99% from Sigma Aldrich and distilled and deionized water); PBS10 (prepared with Na_2HPO_4 99% from Sigma Aldrich; $NaH_2PO_4 \cdot H_2O$ 98% from Sigma Aldrich and NaCl, ACS reagent, from Vetec (Rio de Janeiro, Brazil).

Author Contributions: Conceptualization, R.S.B., A.L.S.L. and K.S.C.L.; methodology, R.S.B., C.G.M.S., A.L.S.L. K.S.C.L. and M.R.D.; software, R.S.B.; validation, R.S.B. and A.L.S.L.; formal analysis, R.S.B., A.L.S.L. and K.S.C.L.; investigation, R.S.B., C.G.M.S., A.L.S.L. and K.S.C.L.; resources, A.L.S.L., T.C.C.F., and K.K.; data curation, R.S.B.; writing—original draft preparation, R.S.B., A.L.S.L.; writing—review and editing, E.N., T.C.C.F. and

K.K.; visualization, E.N., R.S.B.; supervision, A.L.S.L. and K.S.C.L.; project administration, A.L.S.L. and K.S.C.L.; funding acquisition, A.L.S.L., T.C.C.F., and K.S.C.L.

Funding: This research was funded by Military Institute of Engineering, Brazilian financial agencies Conselho Nacional de Pesquisa (CNPq) (Grant No. 308225/2018-0) and Fundação de Amparo à Pesquisa do Estado do Rio de Janeiro (FAPERJ). (Grant No. E-02/202.961/2017), Coordenação de Aperfeiçoamento de Pessoal de Nível Superior (CAPES) Pró-Defesa, and the Brazilian Army. This work was also supported by Excellence project University of Hradec Králové (UHK).

Acknowledgments: Military Institute of Engineering, Institute of Chemical, Biological, Radiological and Nuclear (IDQBRN) Defense, Army Institute of Biology.

Conflicts of Interest: The authors declare no conflicts of interest.

References

1. Severino, L.S.; Auld, D.L.; Baldanzi, M.; Cândido, M.J.D.; Chen, G.; Crosby, W.; Tan, D.; He, X.; Lakshmamma, P.; Lavanya, C.; et al. A Review on the Challenges for Increased Production of Castor. *Agron. J.* **2012**, *104*, 853. [CrossRef]

2. Mutlu, H.; Meier, M.A.R. Castor oil as a renewable resource for the chemical industry. *Eur. J. Lipid Sci. Technol.* **2010**, *112*, 10–30. [CrossRef]

3. Carvalho Melo, W.; Barreto da Silva Nei Pereira Jr, D.; Maria Melo Santa Anna, L. Ethanol production from castor bean cake (Ricinus communis L.) and evaluation of the lethality of the cake for mice. *Quim. Nova* **2008**, *31*, 1104–1106.

4. Silva, B.A.; Stephan, M.P.; Koblitz, M.G.B.; Ascheri, J.L.R. Influência da concentração de NaCl e pH na extração de ricina em torta de mamona (*Ricinus communis* L.) e sua caracterização por eletroforese. *Ciência Rural* **2012**, *42*, 1320–1326. [CrossRef]

5. De Oliveira, A.S.; Campos, J.M.S.; Oliveira, M.R.C.; Brito, A.F.; Filho, S.C.V.; Detmann, E.; Valadares, R.F.D.; de Souza, S.M.; Machado, O.L.T. Nutrient digestibility, nitrogen metabolism and hepatic function of sheep fed diets containing solvent or expeller castorseed meal treated with calcium hydroxide. *Anim. Feed Sci. Technol.* **2010**, *158*, 15–28. [CrossRef]

6. Anandan, S.; Kumar, G.K.A.; Ghosh, J.; Ramachandra, K.S. Effect of different physical and chemical treatments on detoxification of ricin in castor cake. *Anim. Feed Sci. Technol.* **2005**, *120*, 159–168. [CrossRef]

7. Godoy, M.G.; Fernandes, K.V.; Gutarra, M.L.E.; Melo, E.J.T.; Castro, A.M.; Machado, O.L.T.; Freire, D.M.G. Use of Vero cell line to verify the biodetoxification efficiency of castor bean waste. *Process Biochem.* **2012**, *47*, 578–584. [CrossRef]

8. Check Hayden, E. The quick facts about ricin. *Nature* **2013**. [CrossRef]

9. Chemical Weapons Convention|OPCW. Available online: https://www.opcw.org/chemical-weapons-convention (accessed on 20 January 2019).

10. Biological Weapons—UNODA. Available online: https://www.un.org/disarmament/wmd/bio/ (accessed on 24 January 2019).

11. Spivak, L.; Hendrickson, R.G. Ricin. *Crit. Care Clin.* **2005**, *21*, 815–824. [CrossRef] [PubMed]

12. Musshoff, F.; Madea, B. Ricin poisoning and forensic toxicology. *Drug Test. Anal.* **2009**, *1*, 184–191. [CrossRef]

13. Audi, J.; Belson, M.; Patel, M.; Schier, J.; Osterloh, J. Ricin Poisoning. *JAMA* **2005**, *294*, 2342. [CrossRef] [PubMed]

14. Gupta, R.C. *Handbook of Toxicology of Chemical Warfare Agents*; Academic Press: Cambridge, MA, USA, 2009; ISBN 9780128001592.

15. Olsnes, S. The history of ricin, abrin and related toxins. *Toxicon* **2004**, *44*, 361–370. [CrossRef]

16. Olsnes, S.; Kozlov, J.V. Ricin. *Toxicon* **2001**, *39*, 1723–1728. [CrossRef]

17. Dai, J.; Zhao, L.; Yang, H.; Guo, H.; Fan, K.; Wang, H.; Qian, W.; Zhang, D.; Li, B.; Wang, H.; et al. Identification of a novel functional domain of ricin responsible for its potent toxicity. *J. Biol. Chem.* **2011**, *286*, 12166–12171. [CrossRef] [PubMed]

18. Stirpe, F. Ribosome-inactivating proteins. *Toxicon* **2004**, *44*, 371–383. [CrossRef]

19. May, K.L.; Yan, Q.; Tumer, N.E. Targeting ricin to the ribosome. *Toxicon* **2013**, *69*, 143–151. [CrossRef] [PubMed]

20. Hartley, M.R.; Lord, J.M. Cytotoxic ribosome-inactivating lectins from plants. *Biochim. Biophys. Acta Proteins Proteomics* **2004**, *1701*, 1–14. [CrossRef] [PubMed]

21. Doan, L.G. Ricin: Mechanism of Toxicity, Clinical Manifestations, and Vaccine Development. A Review. *J. Toxicol. Clin. Toxicol.* **2004**, *42*, 201–208. [CrossRef]

22. Marconescu, P.S.; Smallshaw, J.E.; Pop, L.M.; Ruback, S.L.; Vitetta, E.S. Intradermal administration of RiVax protects mice from mucosal and systemic ricin intoxication. *Vaccine* **2010**, *28*, 5315–5322. [CrossRef]

23. Yermakova, A.; Klokk, T.I.; O'Hara, J.M.; Cole, R.; Sandvig, K.; Mantis, N.J. Neutralizing Monoclonal Antibodies against Disparate Epitopes on Ricin Toxin's Enzymatic Subunit Interfere with Intracellular Toxin Transport. *Sci. Rep.* **2016**, *6*, 22721. [CrossRef]

24. Pincus, S.H.; Smallshaw, J.E.; Song, K.; Berry, J.; Vitetta, E.S.; Pincus, S.H.; Smallshaw, J.E.; Song, K.; Berry, J.; Vitetta, E.S. Passive and Active Vaccination Strategies to Prevent Ricin Poisoning. *Toxins (Basel)* **2011**, *3*, 1163–1184. [CrossRef] [PubMed]

25. Vitetta, E.S.; Smallshaw, J.E.; Schindler, J. Pilot phase IB clinical trial of an alhydrogel-adsorbed recombinant ricin vaccine. *Clin. Vaccine Immunol.* **2012**, *19*, 1697–1699. [CrossRef] [PubMed]

26. Dubois, J.-L.; Piccirilli, A.; Magne, J.; He, X. Detoxification of castor meal through reactive seed crushing. *Ind. Crops Prod.* **2013**, *43*, 194–199. [CrossRef]

27. Sturm, M.B.; Schramm, V.L. Detecting Ricin: Sensitive Luminescent Assay for Ricin A-Chain Ribosome Depurination Kinetics. *Anal. Chem.* **2009**, *81*, 2847–2853. [CrossRef]

28. Hines, H.B.; Brueggemann, E.E.; Hale, M.L. High-performance liquid chromatography–mass selective detection assay for adenine released from a synthetic RNA substrate by ricin A chain. *Anal. Biochem.* **2004**, *330*, 119–122. [CrossRef]

29. Shyu, H.-F.; Chiao, D.-J.; Liu, H.-W.; Tang, S.-S. Monoclonal Antibody-Based Enzyme Immunoassay for Detection of Ricin. *Hybrid. Hybridomics* **2002**, *21*, 69–73. [CrossRef] [PubMed]

30. Poli, M.A.; Rivera, V.R.; Hewetson, J.F.; Merrill, G.A. Detection of ricin by colorimetric and chemiluminescence ELISA. *Toxicon* **1994**, *32*, 1371–1377. [CrossRef]

31. Leith, A.G.; Griffiths, G.D.; Green, M.A. Quantification of ricin toxin using a highly sensitive avidin/biotin enzyme-linked immunosorbent assay. *J. Forensic Sci. Soc.* **1988**, *28*, 227–236. [CrossRef]

32. Sehgal, P.; Khan, M.; Kumar, O.; Vijayaraghavan, R. Purification, characterization and toxicity profile of ricin isoforms from castor beans. *Food. Chem. Toxicol.* **2010**, *48*, 3171–3176. [CrossRef]

33. Kim, S.-K.; Hancok, D.K.; Wang, L.; Cole, K.D.; Reddy, P.T. Methods to Characterize Ricin for the Development of Reference Materials. *J. Res. Natl. Inst. Stand. Technol.* **2012**, *111*, 313. [CrossRef]

34. Despeyroux, D.; Walker, N.; Pearce, M.; Fisher, M.; McDonnell, M.; Bailey, S.C.; Griffiths, G.D.; Watts, P. Characterization of Ricin Heterogeneity by Electrospray Mass Spectrometry, Capillary Electrophoresis, and Resonant Mirror. *Anal. Biochem.* **2000**, *279*, 23–36. [CrossRef]

35. Fultonlb, R.J.; Blakeyb, D.C.; Knowled, P.P.; Uhrs, J.W.; Thorpeq, P.E.; Vitettas, E.S. Article title. *J. BIOL. CHEM. Purif. Ricin AI AZ B Chains Charact. Their Toxic.* **1986**, *261*.

36. UniProtKB/Swiss-Prot. Available online: https://web.expasy.org/docs/swiss-prot_guideline.html (accessed on 25 January 2019).

37. National Center for Biotechnology Information. Available online: https://www.ncbi.nlm.nih.gov/ (accessed on 25 January 2019).

38. Tezotto-Uliana; Silva, P.P.M.; Kluge, R.A.; Spoto, M.H.F. Gamma Radiation in Plant Foods. *Rev. Virtual Quim.* **2015**, *7*, 267–277. [CrossRef]

39. Lee, J.-W.; Kim, J.-H.; Yook, H.-S.; Kang, K.-O.; Lee, S.-Y.; Hwang, H.-J.; Myung-Woo Byun, A. Effects of Gamma Radiation on the Allergenic and Antigenic Properties of Milk Proteins. *J. Food Prot.* **2001**, *64*, 272–276. [CrossRef] [PubMed]

40. Maity, J.P.; Chakraborty, S.; Kar, S.; Panja, S.; Jean, J.-S.; Samal, A.C.; Chakraborty, A.; Santra, S.C. Effects of gamma irradiation on edible seed protein, amino acids and genomic DNA during sterilization. *Food Chem.* **2009**, *114*, 1237–1244. [CrossRef]

41. Funatsu, G.; Kimura, M.; Funatsu, M. Biochemical studies on ricin. XXVI. Primary structure of Ala chain of ricin D. *Agric. Biol. Chem.* **1979**, *43*, 2221–2224. [CrossRef]

42. Ma, X.; Tang, J.; Li, C.; Liu, Q.; Chen, J.; Li, H.; Guo, L.; Xie, J. Identification and quantification of ricin in biomedical samples by magnetic immunocapture enrichment and liquid chromatography electrospray ionization tandem mass spectrometry. *Anal. Bioanal. Chem.* **2014**, *406*, 5147–5155. [CrossRef] [PubMed]

43. Lord, J.M.; Roberts, L.M.; Robertus, J.D. Ricin: Structure, mode of action, and some current applications. *FASEB J.* **1994**, *8*, 201–208. [CrossRef] [PubMed]

44. Wunschel, D.S.; Melville, A.M.; Ehrhardt, C.J.; Colburn, H.A.; Victry, K.D.; Antolick, K.C.; Wahl, J.H.; Wahl, K.L. Integration of gas chromatography mass spectrometry methods for differentiating ricin preparation methods. *Analyst* **2012**, *137*, 2077. [CrossRef]
45. Cardozo, M.; De Souza, S.P.; Dos Santos Cople Lima, K.; Oliveira, A.A.; Rezende, C.M.; França, T.C.C.; Dos Santos Lima, A.L. Degradation of phenylethylamine and tyramine by gamma radiation process and docking studies of its radiolytes. *J. Braz. Chem. Soc.* **2014**, *25*. [CrossRef]
46. Schieltz, D.M.; McWilliams, L.G.; Kuklenyik, Z.; Prezioso, S.M.; Carter, A.J.; Williamson, Y.M.; McGrath, S.C.; Morse, S.A.; Barr, J.R. Quantification of ricin, RCA and comparison of enzymatic activity in 18 Ricinus communis cultivars by isotope dilution mass spectrometry. *Toxicon* **2015**, *95*, 72–83. [CrossRef] [PubMed]

MDPI
St. Alban-Anlage 66
4052 Basel
Switzerland
Tel. +41 61 683 77 34
Fax +41 61 302 89 18
www.mdpi.com

Toxins Editorial Office
E-mail: toxins@mdpi.com
www.mdpi.com/journal/toxins